The CHICAGO WATER TOWER

The CHICAGO WATER TOWER

John F. Hogan

FOREWORD BY MARC SCHULMAN

Published by The History Press
Charleston, SC
www.historypress.com

Copyright © 2019 by John F. Hogan
All rights reserved

Cover photo courtesy of Julia Thiel.

First published 2019

Manufactured in the United States

ISBN 9781467144971

Library of Congress Control Number: 2019950038

Notice: The information in this book is true and complete to the best of our knowledge. It is offered without guarantee on the part of the author or The History Press. The author and The History Press disclaim all liability in connection with the use of this book.

All rights reserved. No part of this book may be reproduced or transmitted in any form whatsoever without prior written permission from the publisher except in the case of brief quotations embodied in critical articles and reviews.

In memory of George Rodecap

CONTENTS

Foreword, by Marc Schulman — 9
Acknowledgements — 13
Introduction — 15

1. "An Abominably Filthy Soup" — 31
2. Mr. Chesbrough's Tunnel — 43
3. Mr. Boyington's Tower — 52
4. Withstanding the Inferno — 63
5. In Good Company — 74
6. Old 98 — 93
7. Carousels in the Lake — 99
8. Through the Years — 106

Epilogue: An Endnote — 117
Bibliography — 121
Index — 123
About the Author — 127

FOREWORD

The Chicago Water Tower is our town square. Every day, visitors and residents walk through Jane Byrne Park and admire the historic structure without knowing its contribution to Chicago history. In *The Chicago Water Tower*, John Hogan tells the story of how the Water Tower and Pumping Station came to be and their impacts on our city for more than 150 years.

Chicago is defined by its bodies of water: the Chicago River and Lake Michigan. How fitting that our most famous landmark is also defined by water. The Water Tower came to be when a visionary engineer, Ellis Chesbrough, and talented architect, William Boyington, came up with a solution to provide Chicago with a safe supply of water while incorporating a classic design. The idea of where to place the Water Tower and Pumping Station turned out to be an exceptional one. Located in the middle of what would later be known as the Magnificent Mile, the tower stands with Lake Michigan to the north and east and the Chicago River to the south.

John Hogan tells the story of how this all came to be. You want to walk through Jane Byrne Park with his book in hand, checking out the Masonic Fraternity Tablet of 1867 that commemorates the laying of the cornerstone for the Water Tower, the centennial of Chicago in 1937, the 1969 centennial of the Water Tower and its designation as the first water landmark and the contributions of DeWitt Clinton Cregier, who rose from city engineer to mayor. This is Chicago's most hallowed ground and has continued to play such an important role in the city's history.

Foreword

Certainly, what makes the Water Tower and Pumping Station even more special is how they both survived the Chicago fire of 1871. John Hogan tells how the fire destroyed more than seventeen thousand buildings, making the Water Tower one of the few buildings in the North Side to survive and the only one that has been preserved to this day.

Surviving the Great Chicago Fire did not guarantee that the Water Tower would be preserved. The construction of the Michigan Avenue Bridge and the widening of Pine Street created another controversy, as it would have been convenient to tear down or try to move the Water Tower so North Michigan Avenue could be straight. In the process of widening the street, thirty-four buildings were demolished and thirty-three were partially destroyed, but the Water Tower survived thanks to many Chicagoans who came to its aid and appreciated what a survivor we had in that historic structure.

With the name coined by Arthur Rubloff, the Magnificent Mile became one of the premier shopping districts in the world, and the Water Tower is its leading public space. Starting in the 1950s, the Water Tower became the centerpiece of the district's holiday celebrations sponsored by the Greater North Michigan Avenue Association (now the Magnificent Mile Association). Stars of the era, including Betty Grable, Carmen Miranda, Hildegarde and Betty Hutton lit a tree and holiday star on the Water Tower. Those celebrations were the inspiration for me in 1992, when I was the founding chairman of the Magnificent Mile Lights Festival, which is now one of the largest holiday celebrations in the country, lighting more than one million lights on the avenue. Each year, the Water Tower hosts Mickey Mouse, our grand marshal, to test the lights in advance of the procession held on the Saturday before Thanksgiving.

Adjoining the Pumping Station on Chicago Avenue is a firehouse, which Potter Palmer advocated for, and which was designed with a castle-like look by city architect E.F. Hermann. Though it was built at a time when horses pulled the fire engine, the building is still actively operated by Engine Company 98 and Ambulance Company 11. It is a very special place for me, as it is just west of Seneca Park and its Eli M. Schulman playground was named in memory of my dad, who operated Eli's the Place for Steak on the site of what is now Lurie Children's Hospital.

In 2019, Jane Byrne Park hosted the twenty-year reunion for Cows on Parade. More than fifteen cows that had been in the record-breaking public art project came together for a month. We had commissioned "Mooving Eli," which was the only cow on roller skates and was eating a slice of cheesecake. It had originally been placed in front of Eli's the Place for Steak,

Foreword

just east of the Water Tower on Chicago Avenue. What a perfect location to celebrate one of Chicago's most popular events.

As the Water Tower and the Pumping Station celebrate 150 years, we look forward to a bright future for the Magnificent Mile and our historic structures. Surrounded by residential towers, luxury hotels, great restaurants, retailers, cultural institutions and parks, the Water Tower is our link to the past and the best of the future in Chicago.

Marc Schulman
Chicago business, civic and philanthropic leader
President of Eli's Cheesecake Company

ACKNOWLEDGEMENTS

The first time I laid eyes on the old Chicago Water Tower—that curiously elegant assemblage of limestone at Chicago and Michigan Avenues—I'd like to say that I was blown away. Such an admission would have made a creative prologue to the words that follow, but I must confess that I have absolutely no recollection of that initial sighting. As entrenched South Siders, our family rarely—if ever—ventured north of the Loop. If my mother wasn't familiar with some location in the city, no matter where, her stock response was, "Oh, that's up on the North Side," as if to suggest that beyond our place there be dragons.

So, when I did become familiar with the tower later in life, I regarded it as an intriguing throwback to an earlier era, not appreciating the significance of this masterwork hiding in plain sight. I'd never sought an answer to the obvious question: What *is* this thing, and why is it here? Looking at it one day, after many years, stuck in traffic aboard a southbound 146 bus, I decided to find out.

There must be several books about the Water Tower, I reasoned. I looked and looked. I came across a couple of brochures published by the city; a short, typewritten manuscript at the Chicago History Museum; old newspaper clippings; and lots of mentions in Chicago histories, but I couldn't find a book about Chicago's favorite landmark. What follows, for better or worse, is an attempt to fill the void—a benign departure from my previous fare of riots, fires, violent strikes and political corruption.

Like those previous efforts, this one would not have come about without the guidance and encouragement of my editor-in-chief and spouse, Judy Brady.

Acknowledgements

Having her as my wife is reward enough. Her editing and organizational skills, not to mention her eternal patience, represent a bonus.

The Fire Museum of Greater Chicago—one of our area's underappreciated treasures—has long been a source of research support and, perhaps just as important, good old-fashioned camaraderie. The leadership is composed largely of retired members of the Chicago Fire Department whose knowledge of its history and lore seems almost limitless. If a visitor can't get an answer here about firefighting, especially the Chicago way, chances are there is no answer.

Father John McNalis, museum vice president, department chaplain, historian and author, has helped me assemble previous books and came through once again. Father John's four volumes on the history of Chicago firehouses, written with the late Ken Little, are classics.

Retired Chicago firefighter Frank McMenamin, who also happens to be a dentist, was most generous in sharing his photo collection, along with his extensive research on the Great Chicago Fire of 1871.

My lineup of professional contributors is never complete without the name of Andrea Swank, who has now improved the look of yet another of my offerings.

His years as president and CEO of Chicago's Greater North Michigan Avenue Association have given John Maxson a deep knowledge and appreciation of the Water Tower District. John graciously shared many insights that helped guide this book to publication.

Highly accomplished artist Matthew Owens returns to familiar territory. Matt's work has illustrated the pages of several of my previous books. Regardless of subject matter, his renderings continually carry the mark of a talent at the top of his form.

I was honored when Chicago business, civic and philanthropic leader Marc Schulman agreed to write the foreword. The Schulman family has been an integral part of the Water Tower District for years. It's been said that the Water Tower is almost as well known around the country and world as the Schulmans' contribution to palates everywhere with Eli's Cheesecake.

Finally, special thanks to longtime colleague Ben Gibson, senior commissioning editor at Arcadia Publishing and The History Press, who never gave up on this project, even when I felt like giving up myself.

J.F.H
July 2019

INTRODUCTION

It's been called Chicago's most cherished landmark and has been compared to a fairy castle, a medieval fortress and a minaret that would not be out of place in Mecca. Some have mistaken it for a tomb or monument, which in a way, the latter description fits. Ask the security guard sitting at a desk inside the visitors' entrance why the place was built and receive a polite smile and shrug. Most Chicagoans don't give it a second thought. It's just there, like Lake Michigan. During a visit in 1882, Oscar Wilde labeled it "a castellated monstrosity with pepper boxes stuck all over it." As its design was unveiled to a curious public in the spring of 1866, the *Chicago Tribune* foresaw "a handsome looking structure and a decided acquisition to the architectural ornamentation of Chicago." Credit the *Trib* with prescience, but the White Castle hamburger chain won the award for imitation is the best form of flattery when it chose Chicago's Water Tower as the model for its restaurant designs.

Since its completion in 1869, the Water Tower has graced a tidy little park at what is now 806 North Michigan Avenue on the Near North Side—high-end real estate for a structure whose only functional purpose was to conceal a 138-foot vertical water pipe that was removed in 1978 after becoming obsolete decades before. In 2014, the city council voted to name the park in honor of Jane Byrne—Chicago's first female mayor—a few months before her death at age eighty-one.

Byrne lived in a condo across the street from the tower for many years and was known to have cherished the view. Her daughter, Kathy Byrne, said her mother greatly appreciated the civic recognition while she could still

Introduction

Strollers in Jane Byrne Park pass the Chicago Water Tower (*left*) on a pleasant day in early spring. *From the author.*

smell the roses, so to speak. "My great, great grandfather…lived in that area during the time of the Chicago Fire," Kathy Byrne noted. "And the Water Tower is a survivor, and my mother is a survivor, and Chicago is a survivor…. [My mom] said that whatever the trouble was in the city, whatever the crisis was that was brewing, she could look out and see that Water Tower and say, 'You survived the fire, and there was no city left, and you made it….' [I]t was a great symbol of hope and inspiration to her."

Aldermen agreed that the accolade was "past due for many years." So, why the delay? The one word that explains much about life in Chicago: politics. For many of those overdue years, Richard M. Daley served as mayor. Election opponents in 1983, Daley and Byrne were never the best of friends, to put it mildly. Political observers have speculated about a Daley agenda to erase or at least minimize Byrne's legacy. Even the modest bronze tablet mounted at the park's north entrance is understated:

> *Jane Byrne Park*
> *Dedicated 2014*
> *Mayor of Chicago 1979–1983*

Introduction

The marker that greets visitors as they enter the park from the north. *From the author.*

Usually, such markers are accompanied by a paragraph or two citing the honoree's contributions to civic life. "First Female Mayor," at a minimum, would have been a nice touch, but alas.

The Water Tower is embraced by four gardens—one at each corner of the park. A heavily traveled walkway to the west connects Chicago Avenue on the south with Pearson Street and the tower's equally famous namesake—that cathedral of consumerism—Water Tower Place to the north. In a bow to the tower's nineteenth-century ambiance, a few horses and carriages line up along the street at the west edge of the tower, waiting for tourists and maybe some locals, as well. One of the drivers might show up in a top hat and tails from time to time. There used to be many more carriages, but the city—citing traffic congestion and listening to the concerns of animal rights activists—has tightened restrictions on this romantic bit of nostalgia. The future of the carriage rides remains in doubt. Gone, for sure, is the chauvinistic Chicago Transit Authority (CTA) bus driver who used to approach the Michigan and Pearson stop with the announcement: "Water Tower Place, ladies' paradise."

Back to the park. The Byrne tablet has a lot of commemorative company. For example, the southeast garden features a worn ground marker placed by the American Legion in memory of the surrounding neighborhood's World War I dead. Three additional tablets adorn the tower's west wall, recognizing its 1969 centennial and designation by the American Water Works Association as America's first water landmark; the 1937 centennial of the city; and the contributions of DeWitt Clinton Cregier, city engineer, commissioner of public works and mayor from 1889 to 1891.

INTRODUCTION

Above: Carriages await passengers along the short street west of the Water Tower. The vehicles are not as prevalent as they once were due to tighter city restrictions. *From the author.*

Left: A plaque on the west wall of the Water Tower honors DeWitt Cregier, who played a central role in bringing fresh water to the city and later served as mayor. *From the author.*

Introduction

On the opposite, or east, side of the tower, at ground level, is a Masonic Fraternity tablet that recalls the prominent role of that organization in the cornerstone laying ceremonies on March 25, 1867. A chain-connected stretch of eighteen bollards separates that end of the structure from the southbound lanes of Michigan Avenue. A keen-eyed auto or bus occupant stopped in traffic in the right lane can make out the Masonic designation. Barely passable walking space lies between the bollards and the tower. A sign encourages pedestrians to use the walk on the other side. Most do.

The main floor of the tower—the only interior section open to the public—is unremarkable, which, no doubt, explains why it draws comparatively few visitors. Another explanation might be the absence of signage encouraging passersby to step inside. Probably just as well, because there simply isn't much to see. Since 1999, the space has been saddled with the cumbersome title of the City Gallery in the Historic Water Tower—City Gallery for short. It showcases a small sampling of works by local artists and photographers mounted on walls that wrap around the base of the tower proper. Built into the walls are the carved stone heads of two lions whose mouths spouted water once upon a time. Above the interior entranceway

The cornerstone at the northeast end of the Water Tower. The plaque, which was added years later, honors the Masonic official who presided at the dedication ceremonies on March 25, 1867. *From the author.*

Introduction

Lined up like steel sentries, bollards separate the Water Tower from the southbound lanes of Michigan Avenue. *From the author.*

looms a large stone plaque commemorating the structure's dedication. On the east wall of the enclosure stands a barred gate secured by a sturdy padlock that forebodingly suggests the entrance to a crypt. Beyond the gate—concealed from view—lies a narrow, winding iron staircase (not the original) that leads to the cupola. In a modest way, the multi-windowed

Introduction

Pedestrians pass the public entrance at the west side of the Water Tower. Not many people venture inside, perhaps because of a lack of welcoming signage. *From the author.*

cupola served as a forerunner of its neighbor, the John Hancock Observatory, offering views to the public from what was for years the highest point in the city. Much more recently, a physically fit executive of the district's Greater North Michigan Avenue Association, now the Mag Mile Association, admitted to becoming dizzy while making the 237-step, sixteen-story climb. The view from the top, which few in modern times have been privileged to experience, made the ascent worthwhile, he maintained. His descent was uneventful, unlike that of one Frederich Kaiser, a young German immigrant who leaped to his death at midafternoon on October 21, 1875. An unemployed bookkeeper, Kaiser had been confined to a mental institution the previous year and was said to be despondent about his inability to find a job.

An eerily similar incident happened six years later, on June 14, 1881. Another young German, Hugo Von Malapert, beset with money problems, encountered a countryman about the same age at the tower, and the two struck up a conversation. The other man mentioned that he was having difficulty finding employment. He and Von Malapert climbed the staircase to the top where Hugo wrote the name and address of his employer, a jeweler on Adams Street, on a card and told his new friend to stop there the next day for a clerk's job; he was sure he'd find an opening. The pair started back down, but Von Malapert said he'd left something behind and had to go back. As the other man continued down the stairs, he heard a commotion outside. When he exited, he saw Von Malapert's body sprawled on the ground northwest of the tower.

The opposite side of Michigan Avenue is home to the tower's limestone sibling: the old Chicago Water Works complex, which was dedicated at the same time with much less fanfare. The role of the overshadowed sibling was to be the building's fate for the next 150 years and counting—the

Introduction

Above: The first level of the Water Tower—the only section open to the public—is the home of City Gallery, which features the work of local artists and photographers. *From the author.*

Left: Years ago, this lion's head, like its twin on the reverse side of the gallery, spouted water. *From the author.*

Introduction

Left: The gated entrance to the 237-step winding staircase that leads to the top of the Water Tower. The staircase has long been closed to the public. *From the author.*

Right: The Water Tower with two of its famous neighbors: Water Tower Place (*right, front*) and the 875 North Michigan Avenue Building, formerly the John Hancock Center (*right, back*). *From the author.*

workhorse eclipsed by the show horse. Now known as the Chicago Avenue Pumping Station, the works are still operating, pumping at a capacity of 260 million gallons a day and serving a population of some four hundred thousand people in the central city. The Water Tower, on the other hand, has never contained any machinery. The pair were the only public buildings to survive the Great Chicago Fire of 1871, although the Water Works incurred heavy damage and didn't resume partial service for eight days.

The boxy Pumping Station looks more like a fortress than the slender Water Tower and offers much more to see and do. With a public entrance at 163 East Pearson, the two-story front, or west section, encases the three working pumps while the one-story rear portions were designed primarily for boilers, workshops and coal storage. A one-hundred-foot chimney tops the beginning of the rear expanse. Landscaping on the south extends for six hundred feet east of Michigan. A weather-worn air-conditioning

Introduction

The Chicago Avenue Pumping Station on Michigan Avenue, formerly the Chicago Water Works, sits directly east of the Water Tower. Both locations were declared in service on March 3, 1869. *From the author.*

unit mounted halfway down the north wall seems oddly out of place—an unwelcome intrusion into the nineteenth-century limestone continuity.

A short flight of steps at 163 East Pearson leads into a visitor center, lounge area and small branch of the Chicago Public Library. Off the lobby is the box office and entrance to the Lookingglass Theatre, which acquired the former boiler room from the city in 2003. It is believed to be the only performance space in the world under the same roof as a municipal pumping station.

A few steps beyond the Lookingglass box office an impressive sight presents itself: the Pumping Gallery, which allows visitors to gaze down on the four sturdy gray housings that cover the twelve-hundred- and fifteen-hundred-horsepower electric pumps. Each is named for a former station engineer: Old DePaul, Old Kane, Old Pouliot and Old Sallis. The Gallery is aptly named; it has the feeling of an art gallery—handsome, quiet, appealing, conducive to reflection or even meditation. There are no moving parts visible, and on a typical day there are few, if any, employees or visitors to be seen. The place looks spotless, with gauges, fittings, walkways, railings, the pump housings—everything—buffed to a high sheen like a U.S. Navy vessel or the apparatus floor of a firehouse. The visitor looks

INTRODUCTION

Looking southwest, a corner of the Pumping Station with the Water Tower across the street. *From Library of Congress.*

down on the operating deck through glass partitions that line a pair of right-angle walkways. High above, a broad-peaked skylight completes the general feeling of warmth and tranquility.

Immediately east of the Pumping Station, at 202 East Chicago Avenue, is the venerable home of the Chicago Fire Department's Engine Company

Introduction

Left: The former main entrance to the Water Works at 811 North Michigan Avenue. The entrance is now located around the corner to the north, at 163 East Pearson Street. *From the author.*

Below: Lookingglass Theatre Company moved into a one-time boiler room at the Pumping Station in 2003. A city visitor center and a small branch library occupy the adjacent space. *From the author.*

INTRODUCTION

Largely unknown to the public, the Pumping Gallery offers commanding interior views of the Chicago Avenue Pumping Station. *From the author.*

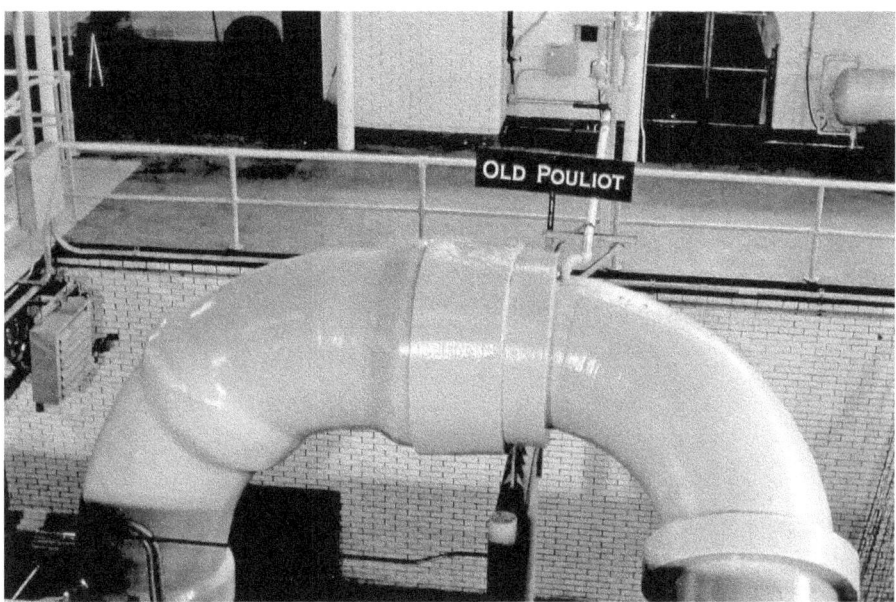

The upper part of one of the four pumps at the Chicago Avenue Pumping Station. Each pump is named for a former station engineer. *From the author.*

Introduction

Left: The ceiling of the Pumping Gallery offers a cathedral-like appearance. *From the author.*

Below: Engine Company 98 and Ambulance Company 11 share quarters at the historic firehouse at 202 East Chicago Avenue. The landmark was built in 1904 to resemble its famous neighbors, the Water Tower and Water Works. *From the author.*

INTRODUCTION

98 and Ambulance 11. This is one of the oldest operating firehouses in the city and the only one to enjoy landmark status. From a distance it looks as if it might have been built about the same time as the Water Tower and Pumping Station, but it didn't arrive until thirty-five years later, in 1904. A close look reveals that the blue-gray Bedford limestone must have come from a separate quarry because it's a slightly different shade than the stone of the more famous neighbors. The firehouse became an official Chicago landmark in 1981.

The Water Tower, Pumping Station and firehouse offer living examples of Old Chicago not merely coexisting with Contemporary Chicago but also enriching a part of town primarily defined by hotels, restaurants and fashionable shopping. Looking back, it's hard to believe that the city—more than once—considered razing the Water Tower for standing in the way of "progress."

1
"AN ABOMINABLY FILTHY SOUP"

The confluence of Lake Michigan and the Chicago River offered an ideal location for early settlers. They could raise crops, trade furs and other goods and transport wares in all directions by canoe or boat. Finding an acceptable source of drinking water became more and more difficult with the passage of time.

When they weren't collecting rainwater in summer or melted snow in winter, people helped themselves to the contents of both lake and river, as did their horses and cattle. In winter, they also chopped holes in the ice and extracted the steaming liquid.

A common well was dug in 1834 near Rush and Kinzie Streets. A few others followed, but many found the water quality poor; "taboo" in the estimation of William Bross, newspaper editor, civic leader and later lieutenant governor of Illinois. So, residents turned to the services of water cart merchants who sold their precious commodity for ten to twenty-five cents a barrel. Most notable among the water salesmen was Anton Berg, who drew his supplies from both bodies of water using a long-handled bucket to fill a large cask. The cask was mounted on a two-wheeled cart pulled by a horse, enabling Berg to make door-to-door deliveries and street sales along the way. Water was emptied from the cask through a short leather hose and into a barrel at the resident's doorstep.

Well water must have been particularly bad for early Chicagoans to prefer river water. The river had not become as polluted as it would in the future with the expansion of its unhealthful secondary function as a receptacle for human waste and storm runoff. Years later, Bross recalled in his best purple prose, "The Chicago River was the source of all the most detestably filthy smells that the breezes of heaven can possibly float to disgusted olfactories." Conditions grew worse as the population increased and expanding stock yards discharged growing amounts of animal waste.

Chicago looked for a more sophisticated approach than delivery carts. In 1836, a group of thirteen prominent citizens formed the Chicago Hydraulic Company to build and operate a waterworks that would draw supplies from the lake. However, an economic downturn in 1839 interfered. It wasn't until the following year that Chicago Hydraulic's backers could raise the $24,000 necessary to start work. The enterprise, completed in 1842, consisted of a twenty-four-horsepower steam engine at the foot of Lake Street that pumped water from a 14-inch diameter wooden pipe situated 150 feet from shore. The water filled a 1,250-barrel reservoir in less than an hour before beginning its journey to the homes and businesses of customers, but it operated only nine hours a day and never on Sunday, unless there was a fire. Supplies were conveyed through a succession of bored-out cedar logs that were 10 feet long and buried 3 feet below ground. The City of Chicago entered into a contract with Chicago Hydraulic to provide service. Rates were set at $10 a year for families and up to $500 annually for large manufacturers. The network eventually spanned nine miles by the time Chicago Hydraulic was superseded by a municipal system.

Despite the breakthrough in water delivery, the system was plagued by frequent malfunctions that forced Chicagoans to fall back on the two old reliables: water taken directly from the river or lake, or a combination of both, supplied by traveling merchants. They could also sample something stronger or go thirsty.

The lake was not immune to some serious problems. During storms, it became excessively muddy. In spring and early summer, young fish found their way into the reservoir, and it was impossible to keep them out. Worst of all, any strong south wind would churn up lake currents and mix sewage from the river into the section of the lake tapped by Chicago Hydraulic's intake system. The result, according to editor Bross, was that "the water from the river, made from the sewage mixed with it into an abominably filthy soup, was pumped up and distributed through the pipes alike to the poorest street gamin and the nabobs of the city." Consumers suffered a comparable experience with the small fish. "It was no uncommon thing," Bross said, "to find the unwelcome fry sporting in one's wash bowl, or dead and stuck in the

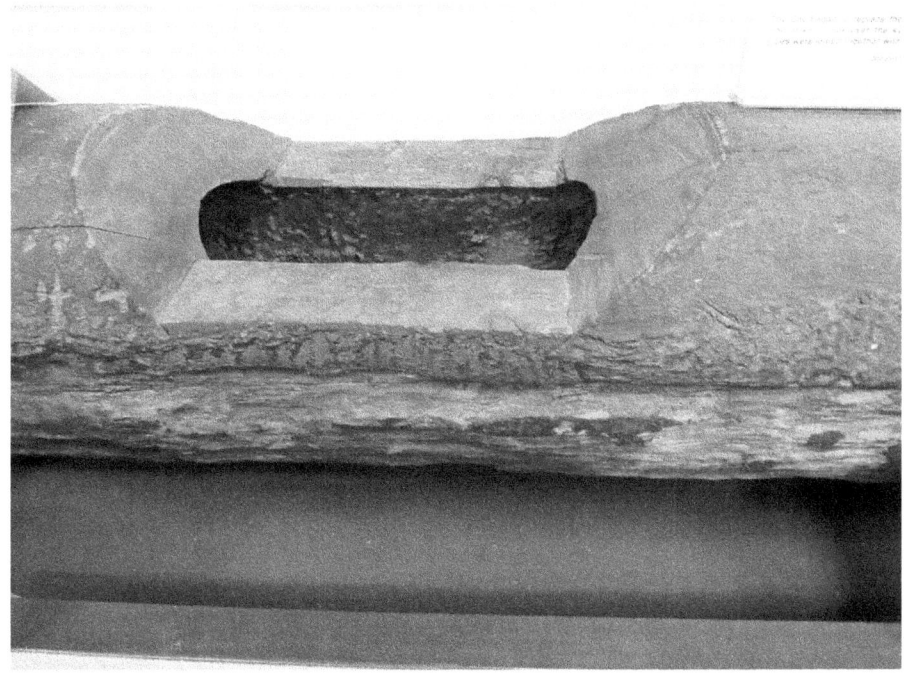

Cutaway of a bored-out log that once conveyed water from a reservoir to the homes of customers. Families were charged ten dollars a year for the service. *From Fire Museum of Greater Chicago.*

faucets. And besides, they would find their way into the hot water reservoir, where they would get stewed up into a very nauseous fish chowder."

There had to be a better way, and in 1851, with the population swelling to thirty-five thousand and buildings growing taller, a new city administration under Mayor Walter Gurnee proposed a municipally owned waterworks, which won overwhelming approval from the voters. The administration hired William J. McAlpine, who had built a waterworks in Albany, New York, to design a new system. After buying out Chicago Hydraulic, the city set about pumping water from six hundred feet offshore into three half-million gallon wrought-iron reservoirs—one for each part of the city—and then into a network of iron pipes that fed the homes and businesses of customers. A new pumping station, located at the foot of Chicago Avenue, went into service in 1854. It became the home of Old Sally, a steam-powered pump capable of producing eight million gallons of water a day. A second pump with a twelve-million-gallon capacity was added in 1857.

The Chicago Water Tower

Right: Chicago's first municipally owned waterworks marked an early attempt to circumvent the pollution problem by drawing lake water from six hundred feet offshore. From *Chicago Public Library*.

Below: The top of this rendering shows the three half-million-gallon reservoirs—one for each part of the city—that were supplied by the waterworks (*front*). From *Chicago Public Library*.

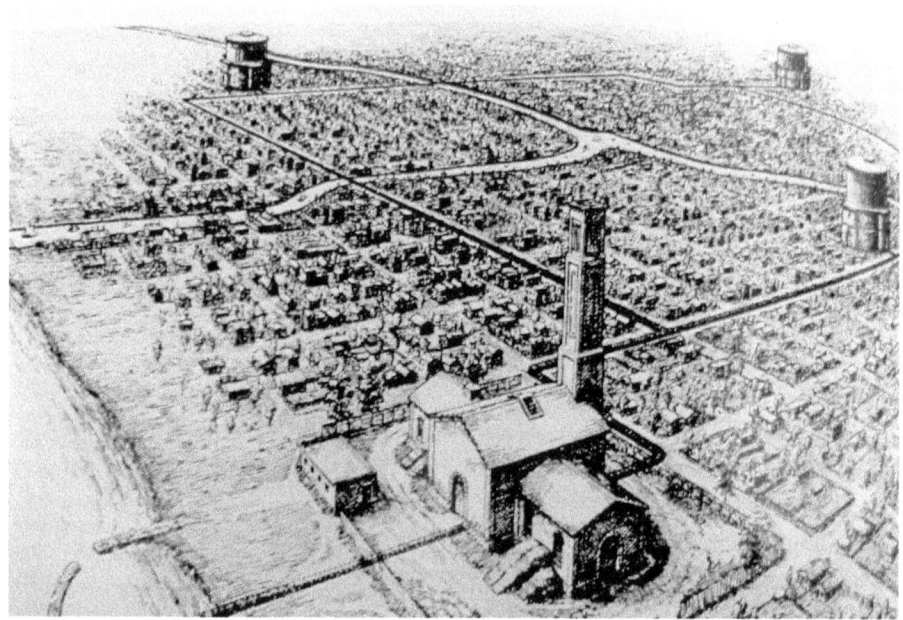

About this time, it was becoming more and more apparent that the problems of water supply and sewage disposal were linked. Chicago historian Joseph Kirkland wrote in 1895, "None of the large cities of the world have such easy access to so vast a body of pure water...and perhaps no other city has committed so many mistakes of engineering in its efforts to preserve the water supply free from the contaminating effects of the city's sewage."

Chicago's biggest mistake was attacking the two problems separately. In early settler days, when the population was comparatively small, sewage disposal was almost an afterthought. Outhouses worked just fine, thank you, but they obviously did nothing to alleviate the accumulations of standing rainwater in low-lying areas. Inhabitants had to confront two incontrovertible facts of nature. First, the flat ground on which their young city rested lay only a few feet above the levels of Lake Michigan and the Chicago River—the natural repositories for runoff. The river was "a sluggish, slimy stream, too lazy to clean itself," in the classic description of author John Lewis Peyton. The second inconvenient truth, as every schoolchild knows, is that water can't flow uphill. The resulting stagnant water created breeding grounds for mosquitoes, foul odors and, worst of all, conditions ripe for the spread of cholera, typhoid, dysentery and diarrhea.

City planners initially tried to eliminate polluted water by creating a spillway of planks along the principal streets leading to the river. The planks, it was hoped, would serve the additional purpose of improving dirt thoroughfares choked with dust or mud, depending on the elements or the time of year. This experiment fizzled when the rains came and floated the planks.

Next, the problem solvers tried digging out the streets to make them lower than the adjoining lots, then paving them with sand and gravel. Rain interfered again, causing the wheels of passing wagons to carry away the mix. As always, such schemes were pushing up against the obstinate river—nearly level with the land and prone to rise, while its already slow, natural eastward current was often further impeded by high lake levels. Chicago found itself trapped in an expanse of putrid water, groping for a way out.

Someone doesn't have to be a doctor or scientist to know that raw sewage, polluted water and abominable odors are incompatible with healthy living. However, by the 1850s, no one had yet drawn the connection between contaminated water and the periodic epidemics of cholera that had plagued the area since 1832. During eight years between 1832 and 1866, when the outbreaks ended, Chicago recorded substantial numbers of cholera deaths. Total fatalities between 1849 and 1856 are believed to

Workers install pavement in 1857 in an attempt to overcome mud and dust that plagued Chicago's streets. A variety of surfaces were tried with limited success. *From Chicago Public Library.*

have surpassed 3,500. The worst year was 1849 with 678 dead—3 percent of the population. Though recordkeeping was imprecise, and cases were frequently misdiagnosed, the city's death rate from all causes was the highest in the nation.

Cholera brings quick and violent death. Its outset is identified by diarrhea, vomiting, cramps and fever, followed by dehydration that causes the face to turn blue. Death usually occurs within a few hours or a day. A hospital that treated victims was described as "a pest house…where cholera patients entered one door, smallpox patients the other, and few left alive," in the words of William K. Beatty, author of a *Chicago History* magazine article.

Beatty cited the recollections of a German immigrant's son:

> *Father drove the hearse and helped lay out the bodies. He used to haul fourteen and fifteen bodies a day and that not during the worse* [sic] *time… He used to work day and night. People died so fast they could not dig holes fast enough. Deaths were so numerous that the coffins and contents were filled up waiting their turn for the holes. Even mother remembers how they were filled up near the old burying ground between Menomenee* [Street] *and North Avenue, now Lincoln Park.*

Medical authorities vaguely associated the cause with poor sanitation, and city officials ordered massive refuse cleanups. Bad nutrition and alcohol consumption were also cited as causes. People were urged to avoid fresh pork, veal and fish, as well as all fruits, vegetables and nuts. Incredibly, a doctor concluded that a woman contracted the disease from eating hot biscuits. No one blamed the water supply, but nonetheless, city leaders were finally beginning to realize that something more had to be done about the sewage problem.

In 1855, the city obtained authority from the state legislature to create a three-member Board of Sewerage Commissioners. Fortunately, one of the members was one of Chicago's shrewdest and most successful early citizens: Chicago's first mayor, William Butler Ogden. The former mayor knew just the person to solve the disposal situation. He was Boston's city engineer, Ellis Sylvester Chesbrough, someone most people today have never heard of but who ranks near the top of the list of those who have shaped Chicago's development.

Ellis Sylvester Chesbrough was brought from Boston to Chicago to deal with the city's persistent sewage and drainage problems. Chesbrough devised a revolutionary plan to draw clean water through a two-mile tunnel beneath the bottom of Lake Michigan. *Rendering by Matthew Owens.*

Chesbrough was a native of Baltimore who was forced to leave school at age thirteen because of his family's declining financial position. Largely self-trained, he worked as a surveyor and construction engineer for several railroads before turning his skills to hydraulics. He became chief engineer for the Boston waterworks and designed the city's system. Whatever Ogden and the board offered must have been attractive because Chesbrough accepted almost immediately. He headed west to tackle the assignment of draining surface water and filth, keeping basements dry and promoting the general healthfulness of the city.

The forty-two-year-old engineer quickly began to justify Ogden's confidence in him. Chesbrough promptly submitted a report that outlined not one but four plans to solve the drainage and waste disposal problem. The effort led to "the first comprehensive sewerage system undertaken by any major city in the United States," wrote Louis P. Cain in *Raising and Watering a City: Ellis Sylvester Chesbrough and Chicago's First Sanitation System,*

"Not one U.S. city at that time had a comprehensive sewerage system, even though most had sewers."

Chesbrough favored the first of his four proposals: drainage directly into the river, which would then carry the sewage into the lake. The other possibilities he raised included drainage directly into the lake, drainage into artificial reservoirs for use as fertilizer and drainage first to the river then to a proposed canal to the Des Plaines River. The last method was adopted at the end of the century but was considered cost-prohibitive in 1855.

Formulating his plan, Chesbrough realized, of course, that he had to solve Chicago's topography problem. In many places, the land was too low to provide a sufficiently steep slope down to the river. Putting sewers under the streets would only exacerbate the situation. How could he surmount nature's obstacle? First, fill in the low places. Chesbrough ran the sewers *on top of* the fill, then covered them with dirt. People laughed. "Don't raise the bridge, lower the river," the man seemed to be saying—or vice-versa.

The chief engineer began building a combined sewer system of brick pipes—three to six feet in diameter—that would accept drainage directly from the streets through manholes placed up to one hundred feet apart and sewage through ducts extending from buildings. The sewer pipes were configured at downward angles that fed the combined mess into the river. Conveniently, the river was being dredged at that time to improve navigation. Dredged soil, hauled away by the wagon load, proved the just right covering for the newly laid pipes. The soil was then topped by brick or cedar blocks, providing the bonus of streets that no longer became seas of mud in the rain or dust bowls in times of drought.

Unfortunately, this process left homeowners and shopkeepers in the literal hole. Vacant lots and buildings had to be raised back to grade level. If not, people who had been living on the ground floor would find themselves occupying English basements. If one neighbor complied with the new order and the family next door did not, the disparity created the need for short flights of wooden steps in the sidewalk out front. Pedestrians continually had to climb up and down to traverse a street. The phenomenon was widespread. Owners were expected to foot the bill for elevating their properties, but not all complied—at least not promptly. Some people flat-out refused. Twenty years passed before the raising of Chicago became complete. At that point, wrote Kirkland, "men's feet were above the place where passed the heads of their predecessors."

Raising wooden cottages was child's play compared with the much more formidable challenge of elevating the larger brick buildings downtown.

The Chicago Water Tower

The uneven process of raising buildings created a disparity that required short flights of steps on wooden sidewalks that forced pedestrians to continually climb up and down. *Rendering by Matthew Owens.*

Also occupying Chesbrough's thoughts was his uneasiness with the end result of his sewage disposal scheme. While the system would rid the city of sewage and drainage and leave basements much drier, he was keenly—perhaps sheepishly—aware of the potential health hazard it created. Both of these items of unfinished business would have to await his return from a unique adventure. Although sewer construction was still underway, the Board of Commissioners handed the engineer-in-chief an ambitious assignment on behalf of a city not yet twenty years old. Perhaps the idea originated with Chesbrough himself, and he sold it to the board. In any event, he was to travel to Europe to compile a report on how the major cities disposed of their sewage.

Ocean travel was becoming easier by the 1850s, but the trip still took about ten days, and Chesbrough made the crossing late in the year. He spent the winter of 1856–57 visiting thirteen cities in England, Scotland and the continent. With varying degrees, he found them confronting problems similar to those back home. In London, for example, Chesbrough observed that all sewage emptied into the Thames. But he hastened to point out that the Thames was six times wider than the Chicago River, had a rapid current and tides that rose to twenty feet. Most sewers in Paris emptied into the Seine, while those of Glasgow discharged into the Clyde. All three cities were working on plans to remove the nuisance, generally focusing on systems involving intercepting sewers. Chesbrough also paid a visit to a town outside Amsterdam to explore its system of using windmills to pump flushing water from a lake into the sewers. He opted to pursue steam pumps.

"None of these great cities," he wrote in his report to the board, "furnishes an exact criterion for keeping sewage out of a river, and yet their experience leads me to fear we may yet, like them, conclude that it will be necessary to keep it out." Consequently, he offered a short-term and long-term solution. Maybe taking a cue from his friends in Holland, he recommended—and the sewage board approved—a plan to flush the river with pure lake water pumped in through the north and south branches. This approach worked reasonably well some of the time, when it wasn't overwhelmed by summer heat, heavy rain or the proliferating discharges from the increasing numbers of stockyards, gas works, distilleries and other industries that used the river as a dump. At such times, the pumping station at Bridgeport on the south branch couldn't move enough lake water to effectively dilute the increased amounts of pollutants.

For the optimal solution, Chesbrough reverted to the fourth option of the plan he presented when he first arrived in town: Build a "steamboat canal" to the Illinois River "that would deliver perfect relief to Chicago…that is, it would furnish a constant and abundant stream from the lake, flowing [south] westerly." Neither Chesbrough nor anyone else used the phrase, but he was proposing nothing less than the reversal of the Chicago River in a way that wouldn't be done successfully until 1900 and would then rank as one of the engineering wonders of the modern world.

Back from his European odyssey, Chesbrough had the opportunity to check on the progress of sewer construction and the raising of the city's buildings. Nobody doubted that the raising would require slow going, but at least it provided spectator sport. "People would gather in the streets by the hundreds," wrote Donald L. Miller in *City of the Century: The Epic of Chicago and the Making of America*, "to watch four- and five-story buildings and entire city blocks at a time—including horse car tracks, lampposts, hydrants and even shade trees—raised as high as twelve feet."

All of these buildings were frame. No one in Chicago had figured out how to raise brick structures, especially larger ones like those found in the central city. Conventional wisdom said the bricks would crack. Luckily, a

Above: A small army of workers turning jackscrews raises a hotel out of the mud without cracking a wall or breaking a glass. As the building rose, fresh landfill was added beneath it. *From Chicago Public Library.*

Left: George M. Pullman traveled to Chicago from New York to raise buildings for a living and stayed to start the railcar manufacturing empire that bore his name. *From Chicago History Museum.*

twenty-eight-year-old New York man named George Mortimer Pullman got word of Chicago's predicament in 1859 and was soon on his way west. Pullman, who operated a house-moving business, had devised a system to raise brick buildings and encountered little trouble finding clients, regardless of the construction type at hand. The Pullman method called for workmen to dig holes into a building's foundation, insert heavy timbers and then place jackscrews on the wood at short intervals around the structure. A small army of men, on command, would turn the jackscrews one revolution at a time and raise the building inch by inch. As it rose, masons would quickly lay fresh landfill beneath it.

Pullman proved that his system worked, but could it handle the tallest brick building in Chicago—the five-and-a-half-story Tremont House? The hotel's owners were hopeful but leery. They had to be tired of watching their guests sit on the front steps and shoot at the ducks that flocked to the gullies along State Street. "I can raise your hotel without breaking a single pane of glass or stopping your business for a day," Pullman assured his prospective clients. Work began in 1861.

The future czar of the railcar empire that bore his name surrounded the elegant building with an incredible army of twelve hundred, who were responsible for turning five thousand jackscrews simultaneously. It took seven weeks to raise the Tremont seven feet and support it with new wooden pilings. Business continued as usual. Not a wall was cracked; not a glass was shattered.

2

MR. CHESBROUGH'S TUNNEL

Chicago's new sewer system was meeting all expectations. The city was cleaner, drier and healthier, but the cost was an unhealthier river and, by extension, an unhealthier portion of the lake that provided the city's drinking water. Sewage and drainage weren't being eliminated, only displaced. Pollutants were emptying into the lake as never before. The existing intake for the city's water supply was a wooden pipe that extended only a few hundred feet into the lake a half mile north of the river's mouth. Chesbrough and members of the water board had known this would happen, and now it was time to do something about it

In 1859, one of the commissioners suggested that the city run an iron pipe one mile out into the lake, beyond the accumulating pollutants. That suggestion was a start, but Chesbrough had a better idea. He became the commissioner of all public works in 1861, reporting to an elected Board of Public Works, which had supplanted the sewage commission. He tried to convince his new bosses that the intake point should lie two miles from shore and be enclosed by a fortress-like structure called a crib, which would be topped by a lighthouse to warn away approaching ships. The crib would come equipped with screens to protect against fish and be staffed all year round by a team of men who would live there to keep the intake gates free of ice floes and floating debris. The most daring aspect of the plan called for the construction of a tunnel more than thirty feet under the lake bottom that would stretch for the entire two miles. Onshore pumps would draw water out of the tunnel and into reservoirs that would feed a network of pipes leading to homes and businesses.

The Chicago Water Tower

Workers tunnel in opposite directions below Lake Michigan. A shaft at the shore (*left*) and another at the Two-Mile Crib formed the entryways for the men and an exit point for the sand and clay they excavated. *Rendering by Matthew Owens.*

Initially the plan was a tough sell. The customary naysayers stepped forward. Digging a tunnel under the lake was crazier than, well, putting sewer pipes on top of existing streets. Despite making an exhaustive study of lake and soil conditions, Chesbrough failed to convince the board. The 1863 election, however, delivered a new, more receptive group of commissioners. Following additional underwater tests, the public works chief got the go-ahead. The board found that his "careful investigation…has satisfied sufficiently to say…we consider it practicable to extend a tunnel under the bed of the lake."

In late summer 1863, the board placed ads in the Chicago and New York newspapers, soliciting bids for the work. The prominent engineering firm Dull and Gowan of Harrisburg, Pennsylvania, was awarded the contract. Though not the lowest bidder, Dull and Gowan won by making a proposal that assumed all risks—not an insignificant guarantee given the dicey nature of the project.

After obtaining the necessary approvals of the state legislature, Congress and the Common (City) Council, the city broke ground for the Chicago Lake

Tunnel on March 17, 1864, St. Patrick's Day. To the cheers of one hundred or so onlookers, Mayor Francis Sherman swung a pickaxe to loosen the dirt that he and other dignitaries shoveled into a wheelbarrow. They watched it be carried away by Colonel James Gowan, one of the contractors. Literally and figuratively, the project was now in the hands of Dull and Gowan. "Every man participating in [the ceremonies was] feeling aware of the great undertaking upon which they were entering, and the disgrace which a failure would bring both upon themselves and the city," according to Jack Wing in 1874 in *The Great Chicago Lake Tunnel*.

Backers of the tunnel plan couldn't be blamed for feeling uneasy. They had become the subjects of ridicule by the eastern press, which scoffed at the idea of building a two-mile tunnel under Lake Michigan to obtain a supply of pure water. It couldn't be done; it shouldn't be done, and Chicago would go bankrupt trying, the argument went. Brushing aside the negativity, Dull and Gowan got down to business.

The contractors began by sinking a brick shaft—nine feet in diameter—a short distance from the shore, where Chicago Avenue met Lake Michigan. The shaft would form the entryway for the men who would dig the tunnel and the exit point for the sand and clay they would excavate. A similar shaft, protected by the crib, was to be sunk at the two-mile point for the crews that would carve out the tunnel from the opposite direction. If the engineers' calculations proved correct—and they were crossing their fingers—the teams would eventually meet thirty feet or so below the bottom of the lake, or about seventy feet from the surface.

Soon after work began near shore, a problem developed in the form of shifting quicksand that defied all attempts to remove it. The contractors temporarily abandoned the brick shaft and substituted an iron cylinder, two and a quarter inches thick and nine feet in diameter, which would extend for twenty-six feet. From that spot, below the quicksand, they reverted to the use of brick.

A journey of a thousand miles begins with one step, and so it must have seemed to the hardy, determined Irish immigrants when they began to sink their picks and shovels into the stubborn blue clay. All digging was done manually at a pace of about fifteen feet a day. More than two and a half years would pass before they would shake hands with their counterparts burrowing from the opposite direction. The laborers had to have considered it a good omen that the groundbreaking took place on St. Patrick's Day.

The sandhogs began their shift through a temporary building atop the shaft that housed a steam engine, office space for the contractors and the

The Chicago Water Tower

A small elevator carried excavators up and down the shafts that led to the tunnels. All digging was done by hand. *Rendering by Matthew Owens.*

requisite tools and materials. An elevator that was powered by the engine carried the men and supplies down the shaft to the point where it connected with the tunnel-in-progress. Given the dimensions of the tunnel—five feet wide and five feet two inches high—only two men at a time could excavate, and they were relieved at regular intervals. Quick on the heels of the miners came the masons who cemented two layers of eight-inch-thick bricks into place. This is how it went without interruption, day and night, day after day—men laboring in dim gas lantern light in space that didn't allow them to fully stand, ever aware of the great inland sea above pressing down and what could happen if something gave way.

Not long after the tunneling got underway, a new problem arose. Air quality was becoming increasingly poor, magnifying the discomfort of working in an environment that was already damp, dark and claustrophobic. The engineers overcame this setback by introducing a steam-powered bellows at the top of the shaft. This apparatus was tied to a perforated tube that ran down the shaft into the tunnel and drew out bad air while forcing in in a fresh supply. As the work moved forward so did the air tube. Problem solved.

With work advancing well beyond the entrance shaft, it became clear that more than manual labor was needed to carry the excavated material back to the elevator. The engineers turned to a method familiar to coal miners: a mini underground rail line. At first, the laborers pushed the mini cars along the rails, but as distances increased, the task was given to a pair of small mules. The quarters were tight, but the animals seemed to catch on easily. They rode the elevator down, just like the men, but didn't come back up each day. In the tunnel they pulled the cars filled with clay back to the shaft, turned around and drew the empties back. About every one thousand feet of the route, men and mules came upon way stations, which were enlarged stopping points carved out for storing material and mixing cement, as well as for stables for the animals and turntables for the cars. These chambers would be bricked up when tunneling was complete.

Back on shore, the time had come for the crib to take a bow following its construction on the North Pier close to the mouth of the river. "Thousands of people were present," historian Wing wrote, "standing upon the house tops, riding upon the river in yawls, and seated in carriages upon the banks and piers." Seven tugboats, including one carrying Illinois governor Richard Oglesby, waited in the harbor to tow the crib to its place in the lake. The mood of the city and country was already upbeat. The date was July 24, 1865, and the Civil War had recently concluded after four bloody years. Wing wrote, "Shortly after ten o'clock, the leviathan moved, rode slowly

The Chicago Water Tower

The Two-Mile Crib under construction onshore. The wooden structure—said to be almost as large as the courthouse—was floated out onto the lake and securely anchored. *From John Carbutt.*

into the river, with streaming flags and the hat of Colonel Gowan swinging over his head from the top. Cannon boomed, hundreds of steam whistles shrieked, bells rang, and thousands of throats cheered lustily." Wing could have been describing the launch of an ocean liner or battleship. All that seemed missing was a bottle of champagne wielded by a female dignitary.

A leviathan it was—said to be almost as big as the courthouse. It was made of one-foot-square timber and tons of iron. It stood forty feet high, ninety-eight feet in diameter and was the shape of a pentagon. Everything was watertight and open at the center to accommodate the shaft that would be sunk to lower the digging crews by elevator, same as their brethren two miles west.

It was a perfect summer day with the lake showing barely a ripple. The crib rode down the skids into the river "as gracefully as any craft ever launched,"

Wing marveled. The tugboats filled with lucky spectators took over from there and churned toward buoys that marked the spot where the engineers' calculations showed a straight line to the shore shaft. Then the bottom of the crib was filled with stone to sink it beneath the water. For good measure, cables were attached and fastened to screws that had been implanted in the blue clay at the lake bottom. The crib was now as immovable as Gibraltar, its developers believed.

The next step called for placement of the iron cylinders that would form the entrance and exit shaft. There were seven, each nine feet long and weighing eleven tons. Getting the immense pipes out to the crib where they would be bolted together presented their handlers with a difficult and time-consuming challenge.

As winter neared, a small house topped by a cupola took shape at the top of the crib, giving it the look of a floating hotel. In fact, the structure was to become a second home for the long-term guests who would dig the tunnel from that end and, later, those assigned to keep the crib in working order. After reporting for duty, it would be months before either group would set foot on land again. A succession of tugboats hurried to get a sufficient supply

Once in place, the Two-Mile Crib became an immovable fortress that was home to the men who would tunnel toward shore and later oversee its operation. *Rendering by Matthew Owens.*

of provisions and construction material to the man-made island before the river, the harbor and probably a portion of the lake froze over. Wing found "something strikingly romantic in the scene....Fifty workmen living an entire winter in a castle built in the lake...a little world within a world. The storms and tempests of a large body of water beating around them in every direction, and they busy digging deeper and farther beneath the lake, as if burrowing out of a prison fortress." The fortress came with its own kitchen and cook, dining room, sleeping quarters and rest and recreation area.

Appropriately, New Year's Day 1866 began with miners at the base of the shaft removing the first shovelful of clay to launch their tunnel leg's long journey westward. The crews burrowing toward them had already advanced more than 4,800 feet from shore. Except for some minor leakage at the last moment, the work progressed smoothly. By late November 1866, the two parties were finally on the verge of linking up within one inch of the center point calculated by the planners. Fears that the separate digs would pass like proverbial ships in the night vanished. Only two feet of clay now separated them—8,275 feet from shore and 2,990 feet from the crib.

On November 30, Chesbrough, James Dull and James Gowan, two supervisors and a group of miners divided themselves into two parties. One, led by Dull, rode a tug out to the crib, descended the shaft and began walking west through the tunnel. The other group, fronted by Chesbrough, headed east on foot from the shore shaft and were first to arrive at the meeting point. Miners chopped through the remaining clay and exposed a continuous tube, two miles long, connecting the crib to the shore. There were cheers and handshakes all around. The scene offered a microcosm of what would happen less than three years later, on May 10, 1869, at Promontory Point, Utah, when the Union Pacific and Central Pacific were joined by the Golden Spike, forming the first transcontinental railroad. Completion of the tunnel eliminated the need for the bellows system that had served the laborers well. As soon as the last clay was removed, pure air filled the tunnel.

One week later, a flag hoisted above the courthouse cupola announced that the time had come for the official ceremonies. Thousands flocked to the lakeshore, but only about two hundred, all men, would get to participate. Invitations had gone out to the aldermen, school board officials and other VIPs for a tour unlike any they had ever experienced. By prearrangement, one group gathered at the State Street Bridge and boarded the tugboat *S.H. Crawford* for the ride out to the crib. A first contingent of the other party descended the shore shaft and boarded railcars that had carried the miners, their tools and materials and the extracted clay. Seated four to a car, none of the passengers complained about the accommodations, though some reportedly made jokes about the mule-drawn trip.

When the train reached the spot a mile and a half out, where the tunneling had finished a week before, Mayor Rice got out of the first car, took a trowel in hand, put the last stone in place and made a little speech in which he called the tunnel "the wonder of America and the world." That group continued to the crib where the VIPs who traveled by tug were already gathered. Soon, a second train arrived carrying the remainder of the passengers, and everyone enjoyed a lunch prepared in the crib's kitchen. The dinner bell, as it were, was the sound of simultaneous cannon fire from crib and shore. Those who came out on the train rode back to shore on the tug. Their counterparts from the boat got their look at the tunnel from the train on the return trip.

The entire project cost the city $464,866.05. Whether any workers were injured or killed during construction, and if so, how many, remains a matter of conjecture. No such records were kept, and even if they were, chances are they would have been destroyed with countless others in the Great Fire four years later. One historian believes the project boasted a near-perfect safety record, while another is convinced that the type of work almost certainly would have caused at least some injuries, if not fatalities. It's unlikely that we will ever know.

3
MR. BOYINGTON'S TOWER

In his masterful biography of Leonardo Da Vinci, Walter Isaacson wrote that the Florence Cathedral, "the most beautiful in Italy…was a triumph of both art and engineering, and linking those two disciplines was a key to [the city's] creativity. This mixing of ideas from different disciplines became the norm as people of diverse talents intermingled." No one would mistake Chicago's Water Works or the Water Tower for Florentine cathedrals, but taken with the crib and the tunnel, they offer a downsized midwestern example of the synergy that Isaacson described. The linkage of the projects combined the talents of two remarkable nineteenth-century minds: Ellis Chesbrough, the engineer, and William Boyington, the artist or—more accurately—the architect.

Five years younger than Chesbrough, William Warren Boyington was another easterner who saw opportunity in the growing young city by the lake. Born in Southwick, Massachusetts, Boyington first worked as a carpenter before moving to New York, where he became an architect and served in the New York State Legislature before heading to Chicago in 1853 at age thirty-five. He would go on to rank as one of the most prolific architects in Chicago and Illinois history. Early on, he designed the old state prison at Joliet in 1858 and the entrance to Chicago's Rosehill Cemetery in 1864. Both are still standing, and Boyington is buried at Rosehill. The prison, long closed, was opened to public tours in 2018. For each, Boyington used the same castellated Gothic style and Joliet-quarried limestone that became his trademarks. His list of credits was so extensive that he once boasted that if all his buildings were

William Warren Boyington, the prolific architect who designed the Water Tower and many other prominent structures. *Rendering by Matthew Owens.*

placed side by side, they would span twenty-five miles. He designed churches, hotels, office buildings and railroad stations, including the Second Presbyterian Church, the first Sherman House Hotel, the original La Salle Street Station and the old Chicago Board of Trade Building. Beyond Chicago, his work extended across the Illinois and Michigan Canal Corridor. Prominent among these jewels is the imposing Hegeler-Carus Mansion in La Salle, built in 1874 and still attracting architecture connoisseurs. When the City of Chicago called on Boyington to crown Chesbrough's engineering marvel, it enlisted a man at the peak of his creative powers.

While the tunnel was still under construction, Boyington was hard at work on his designs for the Water Works and Water Tower. The Board of Public Works, which was overseeing the entire project, understood the desirability of constructing the new Water Works before the tunnel was completed so it would be ready to receive the flow on day one. However, the board felt that the challenge presented by the underwater dig was as much as it cared to tackle at one time. It proved to be a wise decision, although it meant a delay of several months in the regular flow of water.

Even after the meeting of the two tunneling squads, work remained. Sections still had to be bricked up, air pipes removed, the entire tunnel cleaned, intake gates readied and grating placed to keep out floating debris. In addition, planners discovered that alterations were needed to locate the new Water Works on the site of the old, as the master plan dictated. The new buildings had to be larger—too large for the existing foundation—to house the machinery required for the works' expanded operations. The major addition was a new engine capable of pumping eighteen million gallons of water a day into a pair of reservoirs, or wells, located forty feet below ground. It was the largest such engine in the country and the brainchild of another impressive figure in the bold reconfiguration of the water system: DeWitt Clinton Cregier, chief engineer of the pumping station and later mayor of Chicago. The engine was built by the Morgan Iron Works of New York City and increased the capacity of the affectionately named Old Sally

The old Board of Trade Building—one of Boyington's signature accomplishments—was destroyed by the Great Chicago Fire. *From Chicago History Museum.*

and the second engine by three and a half times. Old Sally wouldn't receive her well-earned retirement until early in the next century.

The evolution of hydraulics at that time didn't allow for the basic movement of water from Point A to Point B to Point C—in other words, from the tunnel to the Water Works and into the pipes that supplied the

The Chicago Water Tower

Innovative Water Works chief engineer DeWitt Clinton Cregier also served as mayor of Chicago. *From Chicago Public Library.*

city. Fluctuations in water pressure caused by irregularities in pump cycles—pulsations or surges—could overtax a pipe, causing it to break. Something was needed to relieve undue pressure. The accepted engineering solution called for a vertical standpipe to be connected to the horizontal one. If pressure became too great, water would rise in the standing pipe and relieve the strain on its horizontal counterpart. Think of mercury rising in a thermometer at an accelerated pace. The bottom section of the standpipe had six openings to which the water mains were connected.

Chesbrough and his colleagues realized that the standpipe would have to be tall to accommodate the increasing amounts of water they would be drawing to slake the thirst of a growing city. They settled on 138 feet. But who wants to look at a homely fourteen-story black pipe sprouting like a supersized maypole a short distance north of city center? Enter William W. Boyington with his castellated Gothic trademark, which in plain language meant the incorporation of battlements to create the look of a medieval fortress. It's not difficult to imagine archers taking aim from the narrow windows. The style was quite popular at the time but fell out of favor twenty years later.

Right before his tunnel was placed in service on March 25, 1867, an embarrassing thing happened to Ellis Chesbrough. He had gathered three reporters and set out on a final inspection of the project. With the tunnel partly filled, according to author Donald L. Miller, the four shoved off in a leaky flat-bottom boat. The men stood up in the craft, pushing on the tunnel's sides to propel it forward. On their return to the crib, the boat capsized, plunging the men and their miner's lanterns into cold water up to their necks. Unable to climb back into the boat in the pitch-black tunnel, they had to walk back to reach the ladder at the crib, dragging the boat behind them. Otherwise, the inspection must have proved successful because the floodgates at the crib were reopened and the water roared "like an infant Niagara," in the words of one of the reporters.

The public works chief barely had time to change into dry clothes, assuming that was the case, before he was due at the courthouse for the start of ceremonies marking the cornerstone placement for the Water Tower. The procession that kicked off the day was a half hour late, but if the delay was the result of Chesbrough's mishap, it went unrecorded. He and the other participants and spectators needed their warmest clothes. The day was cold and gray, and the streets were filled with mud. Light snow had fallen earlier. The calendar might proclaim late March as the start of spring, but Chicagoans know better. Still, people turned out by the thousands, shivering and jostling for position—drawn not only by the festivities but also in anticipation of drinking pure water at long last. "From the heart of the lake," noted a reporter for the *Chicago Republican*, "[it] was pouring into the pipes, and…soon the little fishes, the silvery scales, the fears of pestilence, and the other inconveniences of the old system were drawing to an end.… The cornerstone of the Water Works was sinking into its bed, the emblem of the wedding of lake and land."

The dedication ceremonies were a Masonic Order show from start to finish. Cregier was a high-ranking Mason and held several of the organization's top offices, so it's more than likely that he had an influential hand in the orchestration. The big parade might have been led by the police and closed out by the firefighters, but in between strode thousands of Masons of all description. One hundred police officers led the line of march from the courthouse at Dearborn and Randolph Streets to the tower at Pine Street and Chicago Avenue—a distance of about a mile and a half. Far back were the firemen, accompanying a hook-and-ladder, two hose carts and thirteen steamers. A *Chicago Tribune* reporter said the eager horses seemed to believe they had been called out to race to a fire.

Between the public safety cadres, Masons in full uniform of reds, whites, purples and yellows, some sporting white aprons, filled the streets almost as far as the eye could see. Two marching bands separated the ranks at appropriate intervals. There were members of three principal groups: the Grand Council of Royal and Select Masons, the Grand Chapter of the Royal Arch Masons and the Most Worthy Grand Lodge of Free and Accepted Masons. One group was led by a dignitary astride a handsome black charger, another by a grand tyler brandishing a sword. Lesser tylers also wielded swords. Far back came the civic division comprising Mayor Rice, members of the Common Council, other municipal leaders, judges, clergymen and school board members. The press corps was consigned to the rear. Boyington and Cregier marched under the banner of the Most Worthy Grand Lodge of Free and Accepted Masons of the State of Illinois. Boyington, listed as principal architect, carried the symbolic Square, Level and Plumb. Cregier, bearing the title acting deputy grand master, was entrusted with a silver vessel of corn. Chesbrough's name doesn't appear on the roster, so it's likely that he was assigned a somewhat anonymous spot in the civic division as one of the municipal cabinet members.

Before the dignitaries arrived at the site of the tower, which was already under construction, a number of men had mounted a small construction shed to get a better view. The shed collapsed moments before the parade's arrival, depositing "twenty luckless individuals," in the *Tribune*'s words, into an adjacent rubbish pile. No one was seriously injured. The ensuing commotion was dying down when a bare-headed Mayor Rice and others took their places at the platform. The most worthy grand master himself, Jerome R. Gorin of Downstate Decatur, expressed thanks for the honor bestowed on his brotherhood. Gorin was a prominent lawyer, banker and state lawmaker, whose name endures on the metallic tablet affixed to the cornerstone years later.

The grand chaplain offered a prayer; the grand treasurer read a list of articles placed in the cornerstone; the stone was lowered into place by workmen; a choir sang; and the principal architect, Boyington, presented the working tools (Square, Level and Plumb) to the grand master, Cregier, who handed them to the deputy grand master, senior grand warden and junior grand warden. "The stone was consecrated with corn, wine and oil," the *Chicago Republican* added, "the proper addresses being made by the respective officers, The Grand Master struck it three times with a gavel, and the ceremony was over." More speeches followed, the last delivered by

Dedication of the Chicago Water Tower on March 25, 1867. The tower was about 30 percent complete and would go into service almost two years later. *From Frank McMenamin collection.*

Chesbrough "with becoming brevity." Then the VIPs went to lunch, leaving the workmen to resume their labors.

By this time, the tower had already risen taller than 30 feet. It rested on a foundation of 168 piles topped by 7 feet of masonry more than 20 feet below ground. There was no basement. The tower was being built of solid blocks of cream-yellow limestone. It measured 40 feet square at its base, and the first floor was 21 feet tall, with the overall structure tapering gracefully skyward to a height of 154 feet. The spire added another 28 feet. Boyington divided the exterior into five sections with battlement pillars at each of the corners—Oscar Wilde's pepper boxes—and included a balcony around the second section. He placed a doorway with a short flight of steps on each of the four sides at ground level. Only the west entrance remains open.

At the center of it all stood the *raison d'être*: the 138-foot standpipe, 3.5 feet in diameter and surrounded by the winding iron staircase. Small alternating windows were placed along the 237-step ascent to admit natural light. No ceremonies marked the placement of the capstone on August 26, 1867. Boyington topped his masterpiece with a cupola featuring a copper roof with a spire that rested above a circle of large windows. The roof began to take on a greenish hue over time. Undoubtedly, the architect's wisest innovation—some might call it a premonition—was to build the tower of stone, brick and metal. In a word, it was fireproof. Thus the Water Tower and the similarly constructed Pumping Station would be the only public buildings to survive the Great Chicago Fire of 1871.

Across Pine Street, work also continued on the city's more functional but less aesthetically pleasing contribution to the revolution in hydraulic engineering. It's one thing to state that the pumping complex extended six hundred feet east of Pine, but a better perspective can be gained by mentioning that the lake came within fifty feet of the east wall at that time.

The Water Tower and Water Works in 1870—one year after completion and one year before the Great Chicago Fire. *From Chicago History Museum.*

Left: The former main entrance of the Chicago Water Works, now the Chicago Avenue Pumping Station, at 811 North Michigan Avenue. *From the author.*

Right: A one-hundred-foot-tall chimney rises above the rear—or lower—section of the Chicago Avenue Pumping Station. *From the author.*

In the future, three or four blocks of landfill would encroach on the waters and ultimately create space for the Museum of Contemporary Art, Lake Shore Park and Lake Shore Drive.

By late 1866, the components of the new pumping works began falling into place. The engine room foundations were completed in November in anticipation of the arrival of Cregier's colossal brainchild, which was still being built by the Morgan craftsmen in New York. Cregier's engine would join Old Sally and her prosaic, unnamed partner that had been pumping water all along. The engineers ran into some delays in linking the tunnel with its receiving reservoirs and connecting the mains that ran between the engines and the Water Tower. As a consequence, the Cregier machine didn't go into service until late July 1867, although it was ready sooner. Two days later, scaffolding came down, Old Glory went up and the one-hundred-foot chimney atop the rear section of the complex was declared finished.

Even without his accompanying jewel across the street, Boyington's pumping works stood as a grand achievement. It conformed to a principle that would be set forth by another celebrated architect, Daniel Burnham—form follows function. Front and center, facing Pine Street, stood the main entrance. The space above was divided between drafting rooms and sleeping quarters for the engineers. Below were offices and a large reception room. Giant timbers formed the roof of the main building above the viewing gallery that allowed visitors to look down on the engines. The engine room—like the rest of the place—was grand in scale, measuring 142 feet long, 60 feet wide and 36 feet high. Out of sight, 9 feet below the main floor, water delivery pumps and mains formed the connection with the lake tunnel.

The total expanded system—tunnel, buildings, machinery— "the works" was pronounced complete on March 3, 1869, at a cost of $3,146,383.14. However, the euphoria over pure water was short lived. Pollution from the river was reaching beyond the crib's intakes two miles offshore. Chesbrough had counted on periodic heavy rains to clean the river and eliminate, or at least minimize, foul discharges, but for once he had figured wrong. Just the opposite, storms accelerated the flow of pollutants. Chicago's water began to taste and smell almost as bad as it did before completion of the new system. Disinfection by chemicals remained years in the future.

One of the master engineer's initial proposed solutions to the sewage and drainage problems, the reversal of the river, now reemerged. With the aid of a $3 million bond issue and with permission from the state, Chesbrough launched a project to deepen the bed of the Illinois and Michigan Canal where it joined the river at Bridgeport. His engineers erected a temporary dam, then removed it when dredging was finished. Powered by the pumps at that location, river water flowed into the canal for an anticipated journey into the Illinois and Mississippi Rivers. After a somewhat tentative start, the canal began to flow slowly southward.

Once again, Chesbrough was hailed as a genius, at least by Chicagoans. Communities downstream were not so enthralled. The glacially moving canal water rendered its best imitation of its Chicago cousin—the stream "too lazy to clean itself." The resulting semi-stagnation delivered overpowering smells in towns along the Illinois and Michigan Corridor. One year later, the current stopped altogether. Cholera returned in 1873, while smallpox and dysentery epidemics broke out periodically. The Chicago River would not be successfully reversed until the end of the nineteenth century when one of the greatest engineering feats to that time—construction of the Chicago Sanitary and Ship Canal—was accomplished. The new canal was complemented by a new network of sewers that channeled the city's waste to huge sewage processing plants where chemical treatment was applied, solids were deposited on land and the remaining harmless liquids were sent on their way to the Mississippi River and finally into the Gulf of Mexico. The impact on public health was dramatic. Before the opening of the canal, the annual death rate from typhoid was 65 people per 100,000. After the canal and treatment plants began working in harness, the rate plunged to 1 per 100,000.

Ellis Chesbrough wasn't around to witness this final triumph. He died in 1886 at age seventy-three, but Chicago's modern water distribution and sewage systems are the descendants of his pioneering work in the Civil War era.

4
WITHSTANDING THE INFERNO

The legend of Catherine O'Leary's cow kicking over a kerosene lantern and starting the Great Chicago Fire has largely been debunked by historians. It's true that a broken lantern was found in the ruins of the shed behind the O'Leary cottage at 137 DeKoven Street—present site of the Chicago Fire Department's Training Academy and near the intersection of Roosevelt Road and Jefferson Street. However, the cause of the fire is now believed to have been careless smoking, perhaps by Catherine and Patrick O'Leary's tenant Daniel "Peg Leg" Sullivan, neighborhood boys sneaking a puff or someone discarding a lit cigar. But the legend of the cow lives on in some quarters in keeping with the supposed motto of old-time reporters: "Don't let the facts stand in the way of a good story."

Another legend that persists, one that is more easily dismissed, is that the Chicago Water Tower was the only building to survive the fire. The tower, along with the Water Works across the street, were the only public buildings to remain standing, but two private residences in the burned-out district shared that distinction.

In 1871, Chicago, with a population of 334,000, contained more than forty-four thousand wooden buildings, most constructed of pine. It was the largest wood-built city in the world—a highly combustible metropolis just awaiting the spark. If ignition hadn't occurred in the O'Leary barn, sooner or later, it would have happened someplace else with similar devastating results. The harbingers were already gathering like vultures waiting to descend.

From July 3 to October 9, the city recorded only two and a half inches of rain. During most of that time, the southwest wind, "the haze-laden, the thirsty, the grass-killer, the corn-ripener, the hay fever-breeder, the Western sirocco," in the words of historian Kirkland, blew stiffly and relentlessly. Chicago's undermanned and under-equipped fire department, led by forty-five-year-old Toronto-born Robert Williams, was already being run ragged. In the days leading up to the Great Fire, three major fires occurred on October 4, four the following day and five the day after that. On October 7, the eve of the big one, a fire in a planing mill on South Canal Street spread so rapidly that it destroyed a large swath of the southwest central city. Only a superhuman stand by firemen holding their ground fifteen or twenty feet from the inferno at Adams Street prevented the flames from roaring farther north and morphing into the Great Fire one day earlier. As it was, the fire scorched twenty-seven acres bounded by Adams, Clinton and Van Buren Streets and the South Branch of the Chicago River.

Bone-tired firemen returned to quarters and barely had time to put away their equipment before a lookout in a firehouse watchtower about half a mile south of the O'Learys' home spotted flames rising approximately six blocks to the north. It was around 9:30 p.m. Tragically, the lookout mistook the flames for a flare-up of the previous night's blaze and ordered an alarm that sent the exhausted men racing to the wrong location. Further handicapping the response was the unavailability of three of the department's seventeen horse-drawn steam engines. Two were in the repair shop, while the third had been disabled at the previous night's fire. The department's two newest steamers did not respond immediately because they weren't due on that alarm.

By the time the first companies finally arrived, they found fire raging in three outbuildings, the O'Learys' shed, another shed and a paint shop. The O'Leary cottage was spared. All the while, the strong, arid southwest wind continued to blow flames toward a city built of dry fuel. One observer claimed that the fire moved straight from the O'Leary property to the Water Works, as if intent on destroying a mortal enemy. The flames reached that point eventually, but the path was by no means direct, and it took more than six hours. The fire advanced in multiple shifting directions, sometimes outflanking firemen. Heat combined with wind launched huge, swirling bands of fire three hundred to five hundred feet in the air, according to some witnesses. The flaming objects crashed randomly like unguided missiles. A few reached all the way to the crib—an astonishing two miles out in the lake—but were extinguished by the alert crib keeper.

Building after building, block after block went down—the LaSalle Street financial district, the courthouse, the Palmer House, McVickers Theater, Saint Paul's Catholic Church, Saint James Episcopal Church, a lumberyard on the river stacked with a thousand cords of kindling, half a million feet of furniture wood and three quarters of a million wooden shingles. People fled their homes in fear and panic—most on foot—choking the bridges that led from the central city.

Throngs headed east toward the lake with the fire, riding a shift in the wind, close behind. Many waded into the water.

"Carry it or lose it" became the bywords that meant untold numbers of personal possessions—the priceless along with the sentimental—got left behind. People scooped up what they could, starting with the children, then grabbed housewares, silverware, toys, picture frames and whatever seemed close at hand. One man crammed his pet parrot into a small cage once occupied by his dead canary. Practically everything on wheels was packed with human and material cargo and clogged the streets. Those who could afford to hired coaches and wagons, some driven by men eager to turn disaster into a fast dollar. The price of a ride to the lake or sections of the city away from the fire was $150. Some drivers were known to dump their passengers if someone else offered more money.

Well-to-do residents on the outskirts of the Water Works District joined the exodus. George Rumsey, brother of ex-mayor Julian Rumsey, and his family fled their home through the back door as the flames neared the front, but not before rescuing an oil painting of Mrs. Rumsey's father. The Rumseys' seven-year-old daughter decided at the last moment that she didn't want the smoke and ash to soil the good dress she was wearing. She returned to the house to change into something plainer, leaving the first dress behind to burn. Her five-year-old sister carried away a small Swiss clock. All of the Rumseys escaped unharmed.

For a time, it looked as though the fire would stay south of the river's Main Branch. North Siders turned in for the night with a false confidence that would soon be shattered. Judge Lambert Tree, one of Chicago's most distinguished citizens and another neighbor of the Water Tower and Water Works, left a detailed personal account. He described being awoken around midnight by his wife who said that a large fire appeared to be raging to the south. Judge Tree took a look out the window, then left to check on his office at LaSalle and Randolph Streets. He found the courthouse beginning to burn and several other buildings already fully involved. He mentioned that the wind was blowing like a gale. The judge hurried back to his home on Ohio

The Chicago Water Tower

Right: People grabbed whatever possessions they could before fleeing in panic. *From Chicago Public Library.*

Below: Lake Michigan offered refuge to many escaping the fire. *From Chicago Public Library.*

Street, near State, to prepare his family for possible evacuation. Crossing the State Street Bridge, he noticed that dry leaves had been ignited by sparks but weren't causing any damage. All buildings appeared safe. When he got home, Judge Tree took another look out the window and saw that the North Western Railroad Depot and Wright's Livery Stable immediately north of the bridge were burning. Oil cars on the North Western tracks had somehow ignited, spreading fire to the stables alongside. The flames were now little more than a mile south of the Water Works and were moving fast.

Minutes after spotting the stable fire, Judge Tree noticed that a wooden cottage four blocks north of the stable was ablaze. He recalled, "As far as I could discover, there were no buildings intervening between these two points which had not yet taken fire, but it was one of the characteristic features of the conflagration that isolated buildings would catch fire several blocks in advance of the main body of the flames, from the flying sparks and cinders." This phenomenon repeated again and again, although apparently not everyone understood what was happening until too late. For the time being, the crew operating the Water Works believed that the flames would not reach them. Standing between the oncoming fire and their domain loomed the sprawling Diversey and Lill Brewery and Malt House, a square block of substantial brick buildings immediately south and, like its neighbor, extending from Pine Street all the way to the lake. The complex included the Lill residence.

In the early morning of October 9, no buildings were on fire for several blocks south of the brewery, but the plant was not immune to the advance guard of brands and sparks that had touched off so many other structures. This time, the attack from above struck in a spot considered least likely to endanger the entire operation. Only 80 feet into the lake—775 feet east of Pine—stood a small paint and carpenter shop specifically isolated from the rest of the brewery. A large force of brewery workers converged on the shop after it was hit by falling brands. Still believing that their own workplace was safe, men from the Water Works provided reinforcements. It was no use; the paint shop was consumed by fierce fire. Then, instead of spreading to other brewery buildings at that time, the fire turned on the Water Works.

Assistant Engineer Frank Trautman, Timekeeper S.W. Fuller, Station Hand D.W. Fuller and others stood ready. An incessant shower of sparks had been pelting the vicinity. A company of firefighters ran a line of hose from a hydrant while station personnel armed with buckets of water took up positions on the roof. They quickly extinguished several small blazes at the north end, but soon afterward, D.W. Fuller, who was standing just north of

The Water Works chimney looms above the building's partially collapsed roof. Some service was restored only eight days after the fire. *From Fire Museum of Greater Chicago.*

the main building, spotted a brand about twelve feet long whirling toward the Water Tower. The missile passed the tower but struck the northeast corner of the engine house, setting the roof on fire. Fuller looked at his watch; it was 3:20 a.m. Until then, the works had been humming, pumping at a rate of thirty million gallons a day. Despite some breaks in the pipes, the reservoirs remained full, and it looked as if Trautman and his team could keep the water flowing as long as they could hold the flames at bay.

Construction of the main building's roof has provoked controversy. Beneath a thin layer of slate, the roof was wooden, as was the engine room ceiling—circumstances that historian Kirkland called "disastrous and insane." The official report of the Board of Public Works seems somewhat defensive in the way it emphasizes the slate covering and adds, "There was no exterior woodwork in the cornice or elsewhere." Chief Engineer DeWitt Cregier said the building didn't have a metal roof because its predecessor had a ceiling of corrugated iron that condensed the steam and dropped it onto machinery that became damaged as a result. Cregier concluded that "no [fire] department in the world could have stopped the inferno once it jumped the river."

Disputes about construction materials notwithstanding, the roof ignited, and the fire spread to the interior where it fed on woodwork and created a mass of flames. Almost simultaneously, Lill's Brewery, given a short reprieve a little earlier, caught fire and sent wind-driven flames north to put the finishing touches on the Water Works. Firemen trained a stream on the roof, though it was useless. Trautman and his men remained at their posts until part of the roof caved in and the engines stopped. They barely had time to escape. In *History of Chicago*, Kirkland delivered a harsh and unfair judgment of Frank Trautman and the other defenders of the Water Works. Under the subheading "Failure to Defend the Water Works," he asked,

> *Where was the day force, and, if needful, a hundred or a thousand extra men? It certainly must have been known that the ceiling of the engine room was of wood, as were the doors, floors and window frames. All must have perceived that a fearful exigency was at hand, wherein devotion to the very death might be called for. This Waterworks squad was the forlorn hope, the last reliance of a beaten army, and—it makes an orderly retreat; all hands saved.*

Cregier, who was out of the city at the time, asserted later that death would have been "sudden and inevitable" if his men had remained. Cregier pointed out that three of Lill's workers stayed at the brewery too long and were burned beyond recognition.

Kirkland got it right on one score: the Water Works represented the North Side's last hope. When word reached Fire Marshal Williams that the place was on fire, he couldn't believe it—the facility was more than half a mile northeast of the fire line. Williams wondered how he could fight the fire without water and considered the irony of sitting on the shore of a lake 60 miles wide and 360 miles long that could do him no good. When the chief went to the Works for a firsthand look, he found the roof fully involved and the workers inside scrambling. With the surrounding neighborhood a "sea

Looking northeast from Dearborn Street, the Water Tower (*center*) stands unbowed, as does the Water Works chimney (*immediate right*). From *Fire Museum of Greater Chicago*.

of fire," historian A.T. Andreas wrote, the Works became "an utter wreck. Nothing but the naked walls of the building, and the broken and blackened skeletons of three engines were left to mark the spot from which only a few hours before, flowed millions of gallons of pure water." The machine shop and its contents were described as a total loss. Nearly 3 miles of service pipes were melted or damaged.

After Old Sally and its two partners became disabled, a quantity of water remained in the reservoirs, enough to supply parts of the city and beleaguered firefighters for a time. Once that water was gone, the fire department was out of business, except for what it could draw from the river and lake, which were far removed from the current action. Without even an overmatched fire force to block its way, the inferno continued to push north, taking the homes of the wealthy and those of more modest means alike. Some residents soaked carpets and blankets in cisterns that were not dependent on the Water Works and draped them on roofs and across windows. They used their feet to stamp out stray cinders. Only two privately owned buildings survived in the north fire zone. These were the Mahlon Ogden residence on the present site of the Newberry Library, 60 West Walton Street, and the modest home of a police officer's family farther north.

The Chicago Water Tower

Looking east on Chicago Avenue, the Water Tower, with its distinctive cupola, is the tallest structure. The top of the Water Works chimney stands directly to right. *From Fire Museum of Greater Chicago.*

What made the survival of the Ogden mansion particularly unique was that its three stories were all wood. The grand house stood on an expansive lot that also contained a park and several outbuildings. Mahlon Ogden, brother of the city's first mayor, William Ogden, sent his family away when the fire threatened. He and some friends who were visiting were determined to fight the flames. Relying on a cistern on the grounds, they soaked rugs and placed them on the roof and other exposed areas. They splashed the roof with buckets of water and used brooms to stamp and sweep falling cinders. When the water ran out, they turned to barrels of cider stored in the basement. The wooden sidewalks, fences and some of the trees burned, but the mansion survived. Virtue might be its own reward, but an act of charity by Ogden helped save his home. He had donated the square block south of his property to the city for use as Washington Square Park. The open land provided a perfect fire break.

The last house destroyed belonged to Dr. John H. Foster on Fullerton Avenue, just west of Lincoln Park. The fire had lasted a little more than twenty-five hours and gutted a path about four miles long and one mile wide. More than seventeen thousand buildings were destroyed at a loss estimated at $185 million. Only 120 bodies were recovered, but the death

People gather to inspect the destruction in the Water Tower District on the day after the fire. *From Frank McMenamin collection.*

toll is believed to have been two to three times higher because of the large number of missing.

The Water Tower was blackened but not bowed, at least not entirely. Some of the lighter stones and ornamentations were torn off and caught up with the rest of the aerial flotsam and jetsam in the superheated wind. The foundations were weakened, causing a slight tilt that provoked ill-considered comparisons to a more famous leaning tower. Portions of the exterior were left charred, with their "delicate features obliterated or so blunted as to be unsightly," a municipal report declared. Nothing would be done to repair the damage for forty-two years while the tower stood as "an object of pain to the citizens and as wonder to strangers." Mayor Carter Harrison II, who lived near the tower and passed it daily on his way to city hall, supposedly vowed to make repairs but did nothing until midway through his fifth and final term in 1913. The *Tribune*, in an editorial on some other subject, once observed that Harrison, with his feet propped on his desk, was good at that sort of thing.

By refusing to bow to the inferno, the tower served as a guidepost for refugees seeking the safety of the lake, and after the fire, it assisted those trying to find their way back home, or more accurately, to what used to be home. The Water Works, meanwhile, made an amazing comeback under the guidance of team Chesbrough. More than two hundred machinists and laborers toiled day and night, taking their meals at a cook house erected on site. They repaired the damage to the engine house and rebuilt the machine shop. The newest engine, Cregier's masterpiece, was put back in service only eight days after going down. The middle engine was returned one month after the fire, with Old Sally rejoining the group after a seven-week hiatus. Not a bad comeback for three "broken and blackened skeletons."

So, after an impressively short time off the job, the Works were back in action, providing that most vital necessity out of sight, day after day, without fanfare, as it had done since 1869 and would continue to do until the present time. Across the street, the defiant Water Tower would stand proudly as a monument to Chicago's ability to endure the worst, rise from the ashes and continue its journey to recovery.

5

IN GOOD COMPANY

The Water Tower and Pumping Station have always been particular about the company they keep. In the decades before and after the Great Fire of 1871, the tower and its sibling counted among their neighbors the wealthy and prominent—some of the crème de la crème of Chicago society. During the 1920s, a real estate boom transformed Michigan Avenue into a top-of-the-line office and commercial district while the mansions of the neighborhood became supplanted by upscale hotels and apartment houses. Michigan Avenue underwent another overhaul starting in the mid-1950s and continuing for more than a decade as ever larger and taller buildings asserted themselves. Water Tower Place and the John Hancock Building offer prime examples. The Avenue came a long way from the early nineteenth century when its route was part of a trail along the Lake Michigan shoreline called Green Bay Road, where it was home to a small number of wooden buildings. Green Bay Road became Pine Street in 1830, then Michigan Avenue years later.

Pine Street south of the future Water Works was a warehouse and wholesale district that was expanding northward when the fire intervened. As noted, the Diversey and Lill Brewery and Malt house occupied a square block south of the Works. To the north, a succession of other large breweries extended along the lake. Otherwise, the neighborhood was largely residential. The once aptly named street was now lined with elms.

Coexisting with more modest dwellings, the first mansions began to appear in the mid- to late 1830s. "Residences were larger, set on large

After the fire, Pine Street—south of the Water Tower—supplanted Prairie Avenue as the city's most fashionable residential area. The tower is slightly visible on the far right. *From Chicago Public Library.*

wooded lots, some as large as a whole square," according to fire historian Frank McMenamin in *The Fight for Chicago*. "It was reminiscent of a quaint New England village."

While the Mahlon Ogden residence, built in 1859, survived the fire, the George Rumsey and Lambert Tree estates did not. The Rumsey and Tree families rebuilt and were joined by many socially prominent citizens, among them the Pooles, Leiters, Ryersons, McCormicks, Palmers, Farwells and Perry Smiths. Across the city, real estate sales were booming. Between March and October 1872, sales totaled more than $45 million, then reached spectacularly higher levels in the following two years. During the five years after the fire, some one hundred mansions and large homes were built on the Near North Side. The epicenter of silk-stocking Chicago was shifting from Prairie Avenue on the Near South Side to Pine Street and the general vicinity of the Water Tower. "These were the days of gas lamps, hourglass figures, family albums, top hats and gold-headed canes," John Drury wrote in *Old Chicago Homes*.

Though not as opulent as some of its neighbors, the Joseph T. Ryerson family residence at 615 Rush Street began the generation of grand homes built in the neighborhood after the fire. Like a microcosm of the city itself, the Ryerson mansion rose from the ashes in 1873, occupying the same site as its doomed predecessor, which was built in 1860. Ryerson, who parlayed a single store into one of the nation's leading wholesalers of metal products, sold the residence in the 1880s. A succession of owners followed. The last was a Chicago-based real estate firm that bought the mansion and its 8,900-square-foot lot in 2014 and had it demolished the following year to make way for a steak house.

Immediately north of the Ryerson home stood what architecture historians have called the grandest mansion built during the immediate post-fire era: the thirty-five-room Cyrus McCormick residence. The French Second Empire–style dwelling at 675 Rush Street—at the corner of Erie Street—took four years to build, 1875–79. It was the home of the founder of the McCormick Reaper Works, later International Harvester. A wing of the Louvre in Paris reportedly served as a model. So many of McCormick's relatives built homes nearby that the neighborhood became known as McCormickville. Three generations of the family lived in the mansion until 1945 when it was sold. The building was razed in 1955, and today the site is covered by the back of the 900 North Michigan Building.

A McCormick mansion on Rush Street that survives in apparently excellent condition is the so-called double house built in 1875 for Leander McCormick and his son, Robert H. The Joliet limestone structure at 660 Rush is done in Italianate style and is said to be the most intact McCormick family mansion remaining in McCormickville. The double house was declared a Chicago landmark in 2005.

Cyrus McCormick's daughter-in-law, Edith Rockefeller McCormick, always denied that the mansion at 1000 North Lake Shore Drive was purchased and given to her as a wedding present by her father, John D. Rockefeller. Edith was married to Harold McCormick, Cyrus's son. Regardless of how the couple acquired the forty-one-room mansion in 1896, it had two previous owners since it was built in 1883. Reputed to be the richest woman in Chicago, Edith McCormick lavishly furnished the palace with purchases that included a six-hundred-year-old rug owned by Peter the Great and gilded chairs that belonged to Napoleon. The undisputed grande dame of Chicago society, her parties were legendary. After she and Harold divorced in 1921, Edith continued to live in the mansion until the early 1930s, when family members supposedly pressured her to curb her carefree spending and move into the Drake Hotel down the street. The mansion was torn down in 1955 and replaced by a high-rise apartment building.

The mansions nearest the Water Tower, just around the northeast corner, belonged to the Farwell brothers, John V. and Charles B. The brothers made their fortunes as partners in a dry goods business that was acquired by the

A McCormick family mansion that endures—the double house at 660 North Rush Street. *From the author.*

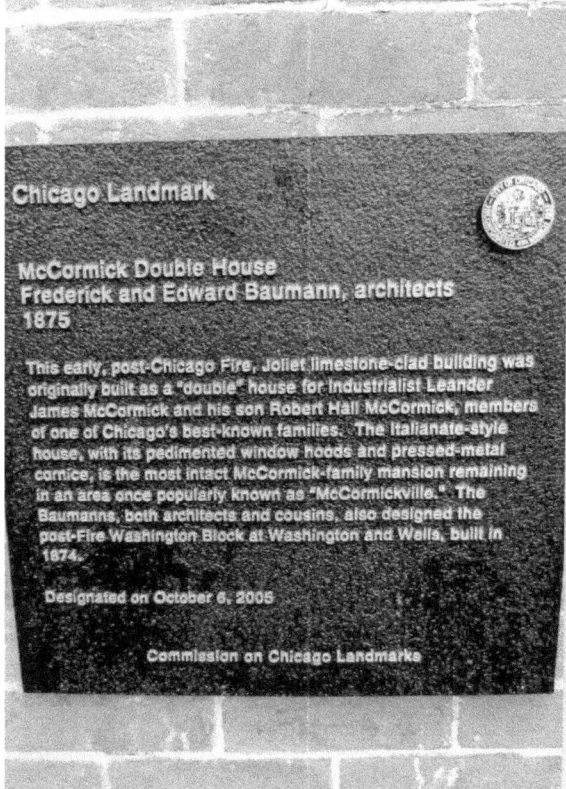

Carson, Pirie Department Store decades later. Their homes stood side by side in the 100 block of East Pearson Street, presenting a "commanding situation" in the description of a guidebook to the 1893 World's Fair. Charles Farwell, also a successful developer, congressman and U.S. senator, built his Queen Anne–style red brick mansion in 1882. Drury admired "the great entrance hall [which] aspired to be the ideal of a baronial manor house with paneled wainscot (of golden oak), an enormous fireplace niche (with a microscopic grate opening), a beamed ceiling, walls and ceilings covered with stenciled canvas, and here and there crossed scimitars, bronze statues, brass plaques, antlers, inlaid tables, Jacobean furniture and two early American Windsor chairs looking very self-conscious and out of place."

John Farwell built an even more elaborate home—one of the most expensive in the neighborhood. It was described at the time as a "turreted baronial castle, quaint and picturesque," with a rich interior of black walnut and cherry. His fortune made, John turned his attention to philanthropy, lending support to evangelist Dwight L. Moody and the YMCA, among other recipients of his largesse. The Farwell mansions were razed in 1946 and were replaced by a women's clothing store that was chicer than the one that occupies the space today.

When Potter and Bertha Palmer built their castle at 1350 North Lake Shore Drive, their goal was to create a mansion to end all mansions. They certainly got their wish and probably their money's worth. At $1 million, they paid a staggering amount for the 1882–84 construction period. The Palmers also chose an unlikely site: an undeveloped area occupied by a patch of frog ponds. But frogs can become princes, and the Palmer mansion was immediately recognized as the largest and most imposing house in Chicago and in the entire country. Designed by prominent architects Henry Ives Cobb and Charles Sumner Frost, the castle measured ten thousand square feet encased in limestone. It was ruled over by the man known as the Father of State Street—founder of a company that later became Marshall Field's and builder of the classic Palmer House Hotel. Every castle needs a queen, and this one had Bertha Honore Palmer, who Drury described as "stately, regal, handsome, wearing her diamond tiara and famous rope of pearls… and presiding over brilliant functions and noble dinners." This is not to mention that she was on the Board of Lady Managers of the World's Fair. In 1950, long after their deaths, Potter and Bertha's castle was demolished. Today, a high-rise apartment building stands in its place.

Another of the Water Tower's neighbors built in the nineteenth century and still standing is the Perry Smith House, at 1400 North Astor Street on

The Chicago Water Tower

The Perry Smith House—a neighbor of the Palmer "castle"—still stands at 1400 North Astor Street on the northwest corner of Astor and Schiller. *From the author.*

the northwest corner of Astor and Schiller. Smith, the heir to a railroad fortune, followed the Palmers into the newly developed section in 1886. He even retained the same architects, Cobb and Frost, who created a red brick Romanesque Revival mansion of seven thousand square feet. A much more recent owner added another three thousand square feet in 1991. The residence includes seven and a half baths and eight fireplaces. It sold in April 2017 for $4 million.

Potter Palmer's plans for what became known as the Gold Coast by no means ended with the construction of his castle. After developing the area with other large houses and nearly three hundred apartments, he bought up almost all the frontage on Pine Street between the Water Tower and Oak Street, according to John P. Stamper in *Chicago's North Michigan Avenue*. There, he built more apartments and townhouses until his death in 1902. Fortunately, Palmer's master plan failed in one significant regard: he was unable to get the Water Tower demolished. In a manuscript on file at the Chicago History Museum, Evelyn S. Nelson points out that the tower "repelled the best efforts of important citizens like the Potter Palmers, Edith Rockefeller McCormick and the Blaines to force its destruction around the turn of the century after they built

A futuristic rendering of North Michigan Avenue completed in 1908—the year before Daniel Burnham unveiled his grandiose Plan of Chicago. *From Chicago Public Library.*

mansions near it." Nelson also cites "complaints by world-renowned architects and artists to the effect that the tower is an eyesore."

These were the waning days of residential development. As early as the 1880s, there had been talk of turning Pine Street into a commercial thoroughfare. Visionaries and opportunists saw the advantages of making Michigan Avenue the primary link between the North and South Sides. The concept received some heavyweight impetus in 1909 with the city's adoption of architect Daniel Burnham's grandiose Plan of Chicago. One of the plan's array of creative features called for the widening and raising of Michigan Avenue to create a "boulevard on stilts." North and south sides of the river would be linked by a new double-deck bridge.

Like many grand plans, this one took years to develop. In 1913, an ordinance to widen Michigan and Pine and build the bridge easily cleared the city council and was signed by Mayor Carter Harrison II. That was the easy part. So, too, was winning approval from the voters to fund the project through three bond issues. Additionally, the city levied two special tax assessments. Much more difficult was the acquisition of the property needed to make room for the widened streets. The widening required the total demolition of thirty-four buildings and the partial demolition of thirty-three more, according to John Stamper. Lawsuits by recalcitrant owners and the city's use of its right of eminent domain kept the project tied up for many years.

At one point, Mayor William H. Thompson's administration entertained the idea of razing the Water Tower or at least moving it to the west. Such a thought was not without some degree of public support. With a tinge of sarcasm, architecture historian Thomas E. Tallmadge attempted to capture the mindset: "With what disdain we regarded the old Water Tower which impinged so inconveniently on Michigan Avenue at Pine Street. It was a symbol of all that was ugly and old-fashioned in the 'Mid-Victorian Age.' It was called 'goldfish bowl architecture' and

Lake Shore Drive at Oak Street in the 1910s. Public pressure convinced the Thompson administration not to relocate or tear down the Water Tower. *From Chicago Public Library.*

was tolerated because it was popularly thought to be the sole building in Chicago that survived the Fire."

The city's street department determined that the tower stood in the way of Michigan Avenue continuing in a straight line and connecting with Lake Shore Drive. "Consultants who were called in," Nelson wrote, "declared that any attempt to remove it would mean its collapse and destruction." All right, said the planners, then we'll have to tear it down. "It was this threat [of] the tower's extinction that really woke up Chicago to an appreciation of its famous landmark. The Chicago Historical Society and civic organizations rose en masse in protest, resulting in the city administration's action to have Michigan Avenue angled to skirt the Tower."

The movement to save the tower got rolling after the president of the Historical Society sat down with the president of the North Center Businessmen's Association. Concurrently, the society was being peppered by requests from members and other concerned citizens to exert its influence to save the tower. At its meeting on October 2, 1918, the society's executive committee adopted a resolution opposing "the removal or destruction of the [Water Tower] as being the only remaining building landmark of Chicago before the Great Fire in the burnt area and that it has endeared itself in the hearts of all the old residents of the city and deserves preservation." A copy of the resolution was sent to Michael Flaherty, superintendent of the Board of Local Improvements, but it was Mayor Thompson who got the message.

Work on the widening and bridge construction began in April 1918, and it ended when the bridge was opened to traffic two years later in May 1920. Stamper said, "It led to the change of the old North Michigan Avenue district from a section of dilapidated warehouse buildings to one of new office buildings and stores, and the conversion of Pine Street into a wide avenue of new office buildings, apartments, hotels, and exclusive shops." With the occupancy changes came a name change. Pine Street was now North Michigan Avenue, the Magnificent Mile or the Mag Mile. Some compared it to the Place de la Concorde in Paris. The price tag came in at just under $15 million.

The imposing Fourth Presbyterian Church on the southwest corner of Michigan Avenue and Delaware Place brought religion to the avenue in advance of the new profit-oriented arrivals. Built between 1912 and 1914, Fourth Pres served gentry who had lived in the area since well before the turn of the century. The Gothic Revival–style house of worship with its 122-foot spire is linked to a parish house, a Sunday school building and a

Auto and pedestrian traffic cross the new Michigan Avenue Bridge with the London Guaranty Building in the background. The corner has changed little in the years since. *From Chicago Public Library.*

gracious courtyard that is set back off the sidewalk and open to the public. The church seems an unlikely presence among the temples of commerce, but it manages to fit comfortably.

Brothers John and Tracy Drake wanted Chicago to have a hotel that would rival the best that New York City could offer. Whether or not they succeeded comes down to a matter of preference, but there remains little doubt that the luxurious palace the brothers named for themselves was the finest in Chicago and continues in the top tier to this day.

The Drake brothers began formulating their plan before the Michigan Avenue overhaul was complete, choosing a site at the northernmost end of the thoroughfare, extending from Walton to Oak Streets. At fourteen stories and eight hundred rooms, the Drake presented a formidable anchor to the end of the avenue. The main entrance on Walton and secondary entrance on Oak were chosen to avoid traffic backups on Michigan. The Oak Street side of the hotel faces a small park and provides unlimited north-facing views of the lake.

The Drake was the first hotel built on Michigan Avenue. The second, the Allerton, is located two blocks south, making the Water Tower the midway point between the two. The Allerton was built between 1922 and 1924. At

The Chicago Water Tower

An overview of the bridge in 2019—taken from an upper floor of the London House, formerly the London Guaranty Building. The lower floors of Tribune Tower are visible in the upper left. *From the author.*

Looking north, traffic flows on a widened Michigan Avenue between the Water Tower and Water Works. The Palmolive Building stands between the two. *From Frank McMenamin collection.*

THE CHICAGO WATER TOWER

The stately Drake Hotel—on Michigan Avenue between Oak and Walton Streets—has retained its top-notch reputation since it became the first hotel built on the avenue in 1924. *From the author.*

701 North Michigan, on the northeast corner of Michigan and Huron, the Allerton took a page from the Drake by putting its main entrance on the Huron Street side. Originally containing more than one thousand rooms, the property was scaled back over the course of several remodels. Since the twenty-three-story hotel's earliest days, iconic signs in large white letters announcing the presence of the long-defunct Tip-Top Tap have adorned both the north and south sides of the building just below the rooftop. The Tip-Top hasn't served a drink since 1961, but its legacy lives on in its signs—an unofficial North Michigan Avenue landmark nearly as recognizable as the Water Tower.

The building at 900 North Michigan held a singular place in Chicago history before it was torn down in 1984 and replaced by a fifty-story tower anchored by Bloomingdale's. It was there, on December 20, 1976, that Mayor Richard J. Daley collapsed and died in his doctor's second-floor office. Prior to that event, the 900's principal distinction was as the only apartment building erected on the avenue in the 1920s (1925–27). It was one of the earliest to offer retail storefronts at street level and apartments on the eight floors above. The building's raised courtyard afforded a view

The Tip-Top Tap hasn't poured a drink since 1961, but its iconic sign on the Allerton Hotel continues to beckon to passersby at 701 North Michigan Avenue. *From the author.*

Loyola University's Lewis Tower has occupied the site at 820 North Michigan Avenue, immediately west of the Water Tower, for many years. The seventeen-story building was originally the home of the Illinois Women's Athletic Club. *From the author.*

of Fourth Pres on the other side of Delaware Place and the Water Tower a block farther south. Residents also enjoyed a view of the lake until taller buildings interfered.

Beginning in the 1920s, Chicago could claim two private women's athletic clubs that were built at about the same time and were two blocks apart. Both offered exercise facilities, swimming pools, ballrooms, dining rooms and space for special events. How they differed might have been in the politics of their leadership.

The first to begin construction was the relative newcomer, the Illinois Women's Athletic Club at 820 North Michigan, immediately west of the Water Tower. The club was founded in 1918 by Mrs. William Severin, part owner of an oil company, leading advocate of Republican Party causes and staunch supporter of women candidates for local political office. (Illinois law allowed women to vote and run for a few offices prior to enactment

The "other" club, the Woman's Chicago Athletic Club, was founded twenty years before its rival, but its new building at 626 North Michigan Avenue wasn't completed until a year after the one up the street. *From the author.*

of national suffrage in 1920.) Mrs. Severin was also an active backer of Mayor Thompson, who was anathema to many mainline Republicans.

Promotional literature described the seventeen-story Gothic building as a monument to the city's largest women's club. Today the building is part of Loyola University's downtown campus and is known as Lewis Tower. The ground floor houses the entrance to the university's Museum of Art. A cornerstone marked "WAC 1925" is the only reminder of the building's previous life.

Billed as the oldest women's athletic club in the nation, the Woman's Chicago Athletic Club was founded twenty years before its counterpart and occupied other locations before it settled at 626 North Michigan. Its members hailed from the city's wealthiest families. In 1898, the club's first president was Mrs. Philip D. Armour, wife of the meatpacking tycoon. The building stands nine stories tall and is topped by a mansard roof, giving it a touch of French elegance. It was completed in 1928—one year after the Illinois club.

The thirty-seven story art deco Palmolive Building at 919 North Michigan was built to house the general offices of the Colgate Palmolive Peet Company. Like other commercial high rises of the era, the building set aside the first two floors for retail establishments with rental office space on the floors not used by Colgate Palmolive. The company courted tenants by touting North Michigan Avenue—in contrast to the Loop—as a place of "quieter streets, clearer skies, more pleasant surroundings." Palmolive's rental brochure, issued at the time of the building's opening in 1928, stated that "modern businessmen, jealous of the hours they now waste yearly in transportation, will welcome this opportunity to work near their homes, live near their offices. Magnificent far-sweeping views, from the upper floors over the lake and city, will bring new pleasure to their business hours, new inspirations to their work." In other words, Colgate Palmolive was selling the sizzle along with the steak.

To call attention to the new building, someone on the Palmolive team, perhaps the inventor himself, conceived the idea of placing a two-billion-candlepower revolving beacon on a one-hundred-foot steel tower atop the structure. The arc would loom six hundred feet above Michigan Avenue and project a beam at least as far as Cleveland, three hundred miles away, and potentially even as far as five hundred miles. The light was a gift of its noted inventor, Edwin A. Sperry, and was named the Lindbergh Beacon in honor of the hero aviator. One ostensible purpose of the beacon was to guide aircraft to Midway Airport. During the war years, 1942–44, the light was turned off to avoid giving assistance to enemy planes. The beacon was shut down again in 1981, this time in response to complaints from residents of newer, taller surrounding buildings. The beacon began its third life in 2007 after it was modified to shift from side to side, facing only the lake.

Playboy Enterprises acquired the Palmolive Building in 1965 and maintained its editorial and business offices there until 1989. Successors renamed it the 919 North Michigan Avenue Building, but the Palmolive name was restored in 2001 by the succeeding ownership.

Arrival of the dazzling Lindbergh Beacon in 1930 provided an ironic twist. The future of North Michigan Avenue, and indeed the entire country, was anything but bright. The depression brought an end to the avenue's boom years that had witnessed the arrival of so many architectural jewels. More than two decades would pass before a revival began.

The Chicago Water Tower

Left: The old and the new. The Pumping Station chimney brackets Water Tower Place to the north. *From the author.*

Right: Looking northwest at the tower, the John Hancock Center can be seen in the upper righthand corner of the frame. *From the author*

When the John Hancock Building made its debut at 875 North Michigan, between 1965 and 1970, it "radically altered the scale of construction on the avenue, tragically exceeding the limits of the 1920s urban context," in the opinion of architecture critic and author Stamper. The one-hundred-story building is set back from the street and separated by a sunken plaza, which also has its critics. The Hancock's dark, sloping walls featuring diagonal bracing suggest a wicker basket design. Located midway between the Palmolive Building and the Chicago Avenue Pumping Station, the Hancock offers a mix of seven hundred condominiums, plus office space and restaurants with retail properties occupying ground level. The building's observation deck, soaring above the avenue on the ninety-fourth floor, competes with its counterpart at Willis (Sears) Tower to woo visitors with the expectation that on a clear day they can see forever, or at least across Lake Michigan.

After the John Hancock insurance company pulled out in February 2018, the building's owners changed its name to the 875 North Michigan Avenue Building. But as anyone associated with Willis Tower can attest, iconic building names die slowly.

The first three floors of Water Tower Place—under construction in 1974—are visible to the left of the Water Tower and Chicago Avenue Pumping Station. Part of the new John Hancock Building can be seen at the far left. *From Frank McMenamin collection.*

The Chicago Water Tower

What Colgate Palmolive sought to do in the late 1920s—lure tenants from the Loop to North Michigan Avenue—Water Tower Place accomplished with shoppers nearly fifty years later. Almost singlehandedly, Water Tower Place shifted the city's retail center of gravity from State Street to the Avenue. In the process, the vertical shopping mall not only adopted but also superseded the name of the venerable landmark. Hop into a cab, say "Water Tower," and guess where you'll end up. A hotel immediately southwest of the tower known as the Water Tower Inn never commanded such universal recognition,

At seventy-four stories, Water Tower Place was built between 1974 and 1976. Sixty-two of its floors rise at the back of a twelve-story common base. The exterior is clad in veined marble. In addition to shops and department stores, the building contains 360 condominiums and a 435-room Ritz-Carlton Hotel.

Not all architecture critics have been enamored of this massive, gray marble presence, but the droves of local shoppers and tourists who flock there all year round obviously couldn't care less.

6
OLD 98

A fter spending part of his fortune populating the Gold Coast with hundreds of apartments and townhouses, and after buying up a large swath of Pine Street, Potter Palmer was more than a little fearful of losing his investments to fire. His original Chicago hotel, the Palmer, had been open for a scant thirteen days when it was destroyed by the Great Fire. Palmer quickly rebuilt and touted his second endeavor as the world's only fireproof hotel.

There was no doubt that for all its wealth, the Gold Coast lacked adequate fire protection. The nearest responders—members of a hose company—were located about half a mile west of the Water Works and more than twice that far from Palmer's Lake Shore Drive mansion. The closest engine companies were still farther away. Streets often clogged with horse-drawn vehicles added to the problem. Whether Palmer alone contemplated this predicament or whether his rich neighbors also harbored concerns is open to conjecture. In any case, the Father of State Street took the lead in lobbying the city to build a firehouse on Chicago Avenue east of Pine. He was said to have sweetened his request with a construction down payment of $10,000, but in doing so, he attached a couple of strings. First, he suggested that the new firehouse be built to resemble his castle. The late Chicago fire historian Ken Little emphatically denied an urban legend that Palmer wanted the firehouse to look like the Water Tower—a logical denial since Palmer and his neighbors allegedly thought the tower was unsightly and wanted it demolished.

THE CHICAGO WATER TOWER

What the Father of State Street thought of the firehouse will never be known because of his death the year before construction started and two years before the facility went into service. Palmer's second request went unfulfilled when the new addition became Engine Company 98. Palmer's address under the old numbering system was 100, and he wanted the same designation for the fire company.

Engine 98's quarters sit atop landfill that displaced the lake waters that once came close to lapping the east end of the Water Works. The fire department acquired the seventy-five-by-two-hundred-foot site through an agreement with the water department. Whether the finished structure more closely resembled Palmer's castle or Boyington's tower would have to be left to the observer. City architect E.F. Hermann gave the firehouse the same castle-like look as the others. Hermann designed a two-story building with a bay for a single horse-drawn engine. A Tudor-style arch spanned the bay. Castellated turrets with mock arrow-slit windows defined the four

Gothic lettering tops the single bay entrance to the quarters of Engine Company 98 and Ambulance Company 11. The firehouse was built in 1904 to resemble the Water Tower and Water Works. *From the author.*

The Chicago Water Tower

Engine Company 98 gallops past the Water Works on slushy Chicago Avenue sometime between 1904 and 1916. *From Fire Museum of Greater Chicago.*

corners. The exterior of the thirty-by-ninety-foot structure was composed of rough-faced, blue-gray limestone placed on top of a smooth-faced stone foundation. Masons cut the stone on site. Adding one more medieval touch, the architect used Gothic script in relief above the bay to spell out "Chicago Fire Department."

Inside, the officers and men outnumbered the horses fourteen to seven when both were at full strength. The animals commanded the rear of the apparatus floor, and the humans shared the second floor with the hayloft until 1916 when the first motorized engine appeared, and the horses were put to pasture. Here and there throughout the building, handsome stained-glass window transoms can be found—an elegant touch befitting the neighborhood, if not necessarily the purpose of the building.

Today, Ambulance Company 11 shares the quarters with Engine Company 98. Together, they protect a part of the Near North Side that includes one of the most heavily traveled pedestrian thoroughfares in the city, along with its high-end retail and residential property. One of Engine 98's members, Lieutenant Edmond Coglianese died of smoke inhalation in 1986 while fighting a three-alarm fire in the budget-class Mark Twain Hotel at 111 West Division Street.

The men of Engine Company 98 pose with their new Ahrens-Fix pumper in 1916. The top of the Water Works with its one-hundred-foot chimney is visible to the left. *From Fire Museum of Greater Chicago.*

Engine 98 awaits a call. Its firehouse is among the oldest in the city. *From the author.*

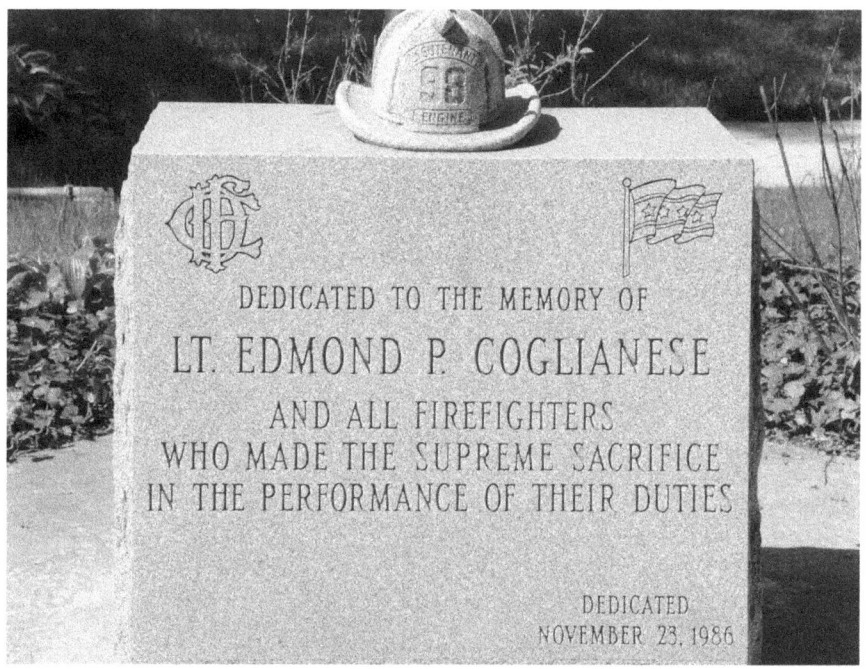

The memory of Lieutenant Edmond Coglianese is preserved by a memorial marker outside the firehouse. Coglianese died in 1986 fighting a hotel fire. *From the author.*

Along the east wall of the firehouse, the joyous sounds of children on swings, slides and jungle gyms can be heard at the Eli Schulman Playground. The space is dedicated to the memory of the legendary restaurateur and creator of world-renowned Eli's Cheesecake. For years, Schulman operated Eli's the Place for Steak across Chicago Avenue from the firehouse. The eatery was a gathering place for celebrities and lesser folk alike and was presided over by "a true Chicago original, a man for all Chicago seasons," as he is described by the bronze marker in the playground. Dedicated in 1990, the playground is separated from grassy Seneca Park by a paving block walkway with benches on each side. The walk connects Chicago Avenue and Pearson Street and, with the park and playground, offers a quiet retreat from Michigan Avenue traffic.

The old firehouse became a Chicago landmark in 1981, but that didn't stop an enterprising restaurant developer from trying to buy the building from the city to convert it into an upscale dining establishment. The plan supposedly enjoyed the support of Mayor Jane Byrne. Once again, a public outcry prevented an affront to Chicago's cherished Water Tower District.

Some two decades later, restaurant investors with better luck and apparently stronger political connections did acquire a landmark Chicago firehouse. The city sold the shuttered quarters of Engine Company 104 at 1401 South Michigan Avenue in the up-and-coming South Loop. The new owners turned the landmark—built in 1905, one year after Engine 98's house—into the Chicago Firehouse Restaurant. Then-mayor Richard M. Daley was a regular patron. Ironically, a serious fire in 2014 forced the eatery to close for twenty-one months before it reopened.

7

CAROUSELS IN THE LAKE

Nineteenth-century historian Jack Wing found "something strikingly romantic" about the city's first water intake crib: "a castle built in the lake…a little world within a world." Chicago's nine past and present cribs have remained objects of curiosity and fascination through the years. A fan of *The Great Gatsby* might detect a comparably romantic touch in the cribs' nighttime beacons as reminders of the light at the end of Daisy's dock. The less curious or fascinated scarcely give the cribs a second or perhaps even a first thought. Like the Water Tower, they get mistaken for other things, such as passing ships, construction barges or lighthouses. In another way, they resemble oversized amusement park carousels, minus the horses—especially the William E. Dever Crib that is two and a half miles off North Avenue Beach, with its candy-striped paint job.

Ellis Chesbrough's Two-Mile Crib—completed in 1867—was so brilliant in concept and execution that it served as the prototype for all eight that followed. The primary purpose of the cribs is to protect the intake vents that draw the city's drinking water from the lake and send it through ten-to-twenty-foot-wide brick tunnels that are buried thirty-two to forty feet under the lakebed, or seventy-five to two hundred feet below the water surface. Water enters the cribs through ports located near their bases, then rises around a central shaft and continues its journey through openings at the top and down to the tunnels.

Only two of the cribs are still active: the Dever and the Edward F. Dunne, 2.3 miles off Fifty-Ninth Street. The Dunne Crib was built in 1909, and the

The Chicago Water Tower

The William E. Dever Crib, the last built (1935), is connected by a footbridge to the Carter Harrison Crib (1900). The Dever is one of two cribs still in service. *From Wikimedia Commons.*

Dever was built in 1935—the last in the series. The all-steel Dever Crib feeds water to the James W. Jardine Water Purification Plant, the world's largest, while the Dunne performs the same function for the Eugene Sawyer Plant, adjacent to Rainbow (Seventy-Ninth Street) Beach on the South Side. Each plant chemically treats the water before sending it through three mains that ultimately lead to a dozen pumping stations. Water then flows into 4,200 miles of pipes on its journey to all parts of the city and 125 suburbs.

Until 1990, the cribs were overseen all year by four-man crews that were rotated weekly. The crib keepers cleared the intake screens of fish, debris and ice. They also handled routine maintenance. Their quarters were simple but comfortable with bunkrooms, a shower, a kitchen and a lounge area with a television set. Some brought barbecue grills and fishing poles.

The two cribs still in operation are automated and monitored by computers and cameras. However, the icebreaker tugboat *James J. Versluis* takes workers out and back every day during winter to deal with ice buildup. If conditions get particularly bad, the crews rely on dynamite. During warm weather, the *Versluis*, which ties up in the slip north of Navy Pier, makes the inspection run about once a week. When a crib requires extended work, such as painting or major repairs, the tradesmen fall back on the living quarters that were once occupied by the crib keepers. Years ago, when a couple of bricklayers were doing a week-long job at the

Entrance to the Jardine Water Purification Plant. The plant treats water drawn from the lake and sends it through a network of mains and pumping stations. *From the author.*

Inspection and maintenance crews are carried to and from the city's two operating cribs by the icebreaker *James J. Versluis*, which ties up north of Navy Pier, alongside the Jardine Water Plant. *From the author.*

Harrison Crib, one of them woke up at about two o'clock in the morning to find a man standing at the foot of his bed. It was a hot summer night, and the workers had left the windows open to catch a breeze. The stranger related that his boat had capsized, but he was able to swim to the crib and pull himself up. The astonished bricklayers gave the sailor coffee and blankets before calling the police marine unit to pick him up.

Such an occurrence would be highly unlikely today. Since 2001, security around the cribs has been tightened significantly. Security systems, including motion detectors, video cameras and door sensors are linked to the Chicago Police Department. A no-vessel zone, marked off by buoys, is in place around each crib and enforced by the police marine unit. The city doesn't like to talk about potential terrorist threats. Some years ago, a spokesman for the Department of Water Management declined to even confirm to a reporter which cribs are operational and where they pump their water.

While the crib keepers are long gone, some of the cribs don't lack for residents. Peregrine falcons have nested and hatched chicks at the Wilson and Sixty-Eighth Street Cribs in recent years.

What follows is the status of the past and present cribs as of 2019:

TWO-MILE CRIB: Following its completion in 1867, Chesbrough's revolutionary Two-Mile Crib off Chicago Avenue stood alone for twenty-six years, supplying the city with water. By the early 1890s, a city that had just passed one million in population found itself in need of substantially greater quantities. That thirst led to the construction of four additional cribs between 1891 and 1900, leaving the Two-Mile Crib to the wrecking crew in 1937.

FOUR-MILE CRIB: For the usual reason—to avoid contamination entering the lake from the river—planners reached ever farther into the lake to locate the city's second crib. The misnamed Four-Mile Crib—it's only 3.3 miles from Monroe Street Harbor—began service in 1891. The stone-and-brick crib was connected to two pumping stations on shore by a pair of tunnels six feet in diameter. An iron lighthouse is mounted atop the structure and displays a white flash every three seconds. The Four-Mile Crib has been out of service for years and is designated for demolition, though no date has been set. The crib is believed to be in good structural shape, which is one reason why the urban conservation group Preservation Chicago advocates its designation as a Chicago landmark and a candidate for reuse.

Along with the Wilson Avenue Crib, also slated for demolition, the Four-Mile fits the definition of *landmark*, according to Preservation Chicago. The Four-Mile and Wilson Avenue Cribs "can't be missed by anyone on Lake

Michigan or the shoreline. Additionally, they are of architectural note and historical importance to Chicago's water supply and engineering history." The group concedes that finding a use for these round, squat structures, miles out in the lake, presents a challenge. It stated, "Perhaps the cribs could be re-purposed as a restaurant, museum, excursion site, [environmental] educational facility, bird sanctuary, or other similar use that can be accessed by boat. If not reused, they should simply be preserved in situ."

Did anyone say gambling casino?

SIXTY-EIGHTH STREET CRIB/EDWARD F. DUNNE CRIB: One year after the Four-Mile Crib entered service, the Sixty-Eighth Street Crib joined the lineup, doing double duty as a lighthouse. The hexagon-shaped brick and cement crib is located 2.3 miles east of Fifty-Ninth Street (well, that's close to Sixty-Eighth Street). While active, the crib supplied water to both sides of the city through a tunnel twenty feet in diameter to the Jardine plant and through a ten-foot tunnel to the Sawyer facility. It came equipped with a navigational warning light atop a steel tower and a bell that tolled every twelve seconds during foggy conditions.

The Sixty-Eighth Street Crib is connected to its partner, the Edward F. Dunne Crib, by a fifty-foot-long steel footbridge, which was added in 1909. The bridge allows maintenance workers convenient access to both locations. The Sixty-Eighth Street side still contains the living quarters once occupied by the crib keepers. Standing in thirty feet of water, the newly completed Dunne Crib was waiting for workers to finish digging its connecting tunnel when a disastrous fire occurred at the nearby temporary construction crib and took the lives of approximately sixty men. The crib is named for the mayor who held office when its plans were adopted.

CARTER H. HARRISON CRIB/WILLIAM E. DEVER CRIB: Named for two more Chicago mayors, the Harrison and Dever Cribs went into service thirty-five years apart: 1900 and 1935, respectively. Like the Sixty-Eighth Street and Dunne Cribs, these two are connected by a footbridge that is one hundred feet long, but only the Harrison comes equipped with living quarters. It was supposed to be phased out with the arrival of its neighbor, but an unanticipated higher demand for water kept both cribs operating until 1997.

Maybe the city was guilty of pressing its luck with the Harrison. The three surviving cribs built in the twentieth century are connected to shore by tunnels bored through limestone and are expected to endure for a millennium. The nineteenth-century models, including the Harrison, followed the original Chesbrough script and tunneled through blue clay. In

1997, an invasion of zebra mussels prompted the city to drain one of the Harrison's tunnels and install a chlorination system that would eradicate the pesky visitors. But not all of the engineers were sold on the idea. Some argued that without the internal pressure of the water, the tunnel would collapse. Others, most notably the outside consultants and contractors, maintained that the tunnel would hold up just fine. The naysayers were proved correct when sections of the tunnel did collapse. The city sealed off the segment of the tunnel that ran under Lake Shore Drive and plugged it with concrete to prevent a repetition with potentially far more serious consequences. Construction snarled traffic on the drive and inconvenienced bikers and other users of the lakefront trail. The repairs cost the city $5.3 million, which it attempted to recover by suing the outside engineering and contracting firms. Roughly twenty years later, the Harrison and Dever Cribs got a $5 million painting, cleaning and repair makeover.

LAKE VIEW CRIB: The Lake View Crib was finally completed in 1896 after the city inherited the construction contract from the Village of Lake View under its 1889 annexation. The intervening years were marked by tunnel, intake and pumping station problems. The short-lived Lake View Crib more closely resembled an actual dwelling than any of the other worker-occupied cribs. This distinction was provided by a succession of ordinary-looking residential windows that wrapped around the second of two stories. Octagon-shaped, the crib stood slightly less than two miles east of the former Montrose Boulevard. The crib's lighthouse—featuring a circular balcony and bell—looked like an actual lighthouse rather than a lamp atop a pole. The Lake View Crib was demolished in 1924 after the startup of the Wilson Avenue Crib rendered it unnecessary,

LAWRENCE AVENUE CRIB: Built in 1915, the Lawrence Avenue Crib by far had the shortest life of any of the nine. Only a few years after it was built, it was abandoned, surrounded by Lincoln Park landfill and supplanted by the Wilson Avenue Crib. The Lawrence Avenue Crib stood only fourteen hundred feet, or less than one-third of a mile, offshore. It rested in just fourteen feet of water but rose twenty-two feet above the water line. The crib had a conical roof. From a distance, its circular concrete base gave the appearance of a stockade fence. A fixed red light, mounted atop a pole, could be seen about seven miles away, according to a War Department Corps of Engineers report.

WILSON AVENUE CRIB: The Wilson Avenue Crib turned one hundred years old in 2018, but there were no fireworks to mark the centennial. Why would there be? From its shaky start, the place has always seemed like

The Chicago Water Tower

Long out of service, the Wilson Avenue Crib turned one hundred years old in 2018—three years after the city announced plans to raze it. No date has been set for demolition. *From Wikimedia Commons.*

a stepchild, a bit lonesome sitting out there, 2.1 miles off the far end of Montrose Harbor. Four of the surviving cribs are paired, while the other, the Four-Mile, has been expanded over the years and presents "a unique Romanesque appearance and is capped with a lighthouse." Once it properly settled into the lakebed after a delay of several years, the Wilson Avenue Crib started supplying water through eight miles of tunnels. Its granite block superstructure rests on top of a ninety-foot steel caisson that encloses a forty-foot diameter inner well chamber. It resembles a plain-looking cupcake with a single candle, that is, light tower. The crib had already been on standby for many years in 2015 when the city announced plans to demolish it.

Every so often baseball fans watching a Cubs game on television get a glimpse of the Wilson Avenue Crib when the director is looking for a creative cutaway shot.

8
THROUGH THE YEARS

Looking over a wide range of images of the Water Tower covering a number of years, a friend mentioned how much the beloved landmark has changed. It really hasn't, of course, it only seems that way through different seasons, years, lighting, cameras, prints or that ultimate judgment factor: the eye of the beholder. A few patch-up jobs notwithstanding, the tower has remained the same for the past 150 years, while the surrounding neighborhood and city have changed in ways that scarcely could have been imagined in 1869. So many historic structures have met the wrecking ball that their photos fill coffee table books. It's comforting to know, as life continues to change at warp speed, that we'll always have—not Paris—Mr. Boyington's gem.

The benchmarks of the tower story are more or less:

> **1834** Chicago digs a common well near Rush and Kinzie Streets.
> **1836** The privately owned Chicago Hydraulic Company is formed to operate a waterworks.
> **1842** Chicago Hydraulic begins supplying the city with lake water pumped through a fourteen-inch diameter pipe extending 150 feet from shore.
> **1851** The City of Chicago buys Chicago Hydraulic and creates a municipal waterworks that builds three reservoirs and connects them to a network of pipes.

1854 A new waterworks goes into service at the foot of Chicago Avenue. Old Sally, a steam-powered pump capable of delivering eight million gallons of water a day, is installed.

1855 The city creates a Board of Sewerage Commissioners, which hires Boston's Ellis Chesbrough as chief engineer. Chesbrough designs the first comprehensive sewer system in the nation to address Chicago's drainage and waste disposal problems.

1864 Ground is broken for Chesbrough's revolutionary lake tunnel—a two-mile pathway under the lakebed that will carry water from an intake crib to onshore distribution facilities.

1865 The intake crib is built on land and towed to its resting place two miles into the lake.

1867 The lake tunnel is placed in service. Architect William Boyington designs a new waterworks complex along with a 182-foot limestone tower to mask the system's vertical standpipe. Tower and water works are formally dedicated with a lavish parade and ceremony.

1869 The Water Tower, water works and all attendant equipment are officially declared complete.

1871 The Great Fire ravages the city. The Water Tower becomes the only public building in the fire zone to survive. The water works is badly damaged but returns to service in a matter of weeks.

1872 The city builds a second tunnel under the lake to double the capacity of the original and to assure more than adequate supplies to the South and West Sides. The new tunnel runs from a point on the lakeshore—about seventy feet south of the old one—and continues a parallel route until it makes a right turn to link up with the crib. During its construction, the crib is fitted with a second well to accommodate an additional tunnel. Workers performing the new excavation have two advantages over previous crews. The tunnel is illuminated by gas, and its larger size allows the men to stand upright in the center without removing their hats.

1875 The brutal winter of 1874–75 offers some hardy Chicagoans a novel but daunting pastime—walking across the ice-bound inner lake to the crib. The lake is frozen for several miles away from shore, well beyond the crib two miles out. A *Scribner's Monthly* rendering shows more than two dozen men standing on the crib deck looking out across the ice, in addition to other men who appear to be trying to make their way across large frozen shards.

Today, unauthorized visits to a crib might get the callers arrested, but in the 1800s, the crib force apparently welcomed the company.

A hiker beginning the trek across the frozen lake found the going fairly smooth at first, but farther out, conditions became much more challenging. Within a half-mile-diameter circle around the crib, broken blocks of ragged ice piled up in clusters eight to ten feet high. Two of the crib's intake gates became completely clogged, leaving only one in operation. Ellis Chesbrough decided to take a first-hand look. He'd been alerted to the problems by the crib keepers who used the recently installed telegraph that linked the crib to the Board of Public Works office.

As he did with tunnel inspections, Chesbrough organized a scouting party. Fifteen men, including a *Tribune* reporter, set out from the Water Works on March 4, 1875, the city's thirty-eighth birthday. The reporter mentioned that he and colleagues had previously visited the crib via tunnel and tugboat but never across a field of ice. He likened the group's preparations to those of a polar expedition "with everyone bearing ropes round their shoulders and boat-hooks in their hands." The half-dozen Public Works employees in the party carried planks, poles, ropes and other equipment.

The original plan called for the use of a twenty-foot boat, hopefully to navigate enough patches of open water. The idea failed when some of the men were unable to dislodge the boat from its winter resting place. The chief engineer didn't seem to have much luck with boats. Undaunted, the group began its march. At age sixty-one, Chesbrough impressed his journalist companion by attempting "a feat which but very few men of his years in the city could accomplish."

The party found an unbroken surface of snow and smooth ice for half a mile before encountering the large slabs. The flat ice in between appeared so clear and glossy that the lead hiker kept tapping it with a pole to make sure it was sufficiently firm. Nonetheless, two of his followers stepped through the surface but came away with only wet feet. By now, a quarter mile separated the men from first to last. If one of the stragglers had gone under, the newsman feared, others might not have been able to save him.

Arriving at the crib forty-five minutes after starting, the visitors were welcomed by the chief crib keeper, who reported that all three gates were now open and receiving water, despite the presence of ice piles fifteen to twenty feet above the water line. The reporter believed these "miniature icebergs" extended all the way to the bottom of the lake.

Apparently still fresh after his walk across the lake, the newsman climbed to the top of the 120-foot lighthouse. All he saw was more ice. "There is a certain grandeur in such a sight," he observed, "but it soon became monotonous."

It took little more than half an hour for the men to make the return trip with the wind at their backs. *Tribune* reporters had now made the excursion on water, below water and across ice. All that was left, this one suggested, was a visit by balloon.

> **1878** The year before Thomas Edison perfects the first successful incandescent electric light bulb, electric light shines from the top of the Water Tower.

Arc lighting had been around for a while by the time Chicago's superintendent of fire alarm telegraphs, John P. Barrett, gave local residents a glimpse of this novelty. The French had been experimenting with arc lighting since as early as the 1840s and used it to illuminate the streets of Paris in the 1870s. Americans got a look at an arc lamp at the Philadelphia Centennial Exposition of 1876.

Alerted by stories in the newspapers, a crowd gathered at the Water Tower to watch Barrett demonstrate the concept in which light is produced by an arc created when current passes over gas between two incandescent electrodes. According to the *Tribune* reporter on hand, "The gleam slowly gathered into a ray, which slanted through the darkness, and succeeded in bringing into unpleasant prominence a cow shed in the vicinity. The ray at its brightest was about a foot broad at the start, and widened to perhaps twenty feet at the base of the ground."

> **1900** The Chicago Sanitary District reverses the flow of the Chicago River by creating the Sanitary and Ship Canal to carry chemically treated sewage away from the city.
>
> **1904** The Chicago Fire Department's Engine Company 98 moves into its new quarters at 202 East Chicago Avenue. The limestone, castle-like firehouse is built to resemble its

famous neighbors, the Water Tower and Chicago Avenue Pumping Station.

1906 Nearly forty years after its installation, the Water Tower standpipe is declared obsolete by city engineers who recommend that the tower and the pipe be eliminated. A public uproar saves the tower from demolition. The standpipe wouldn't be removed for more than seventy years.

1909 The worst disaster in the history of the water system occurs on the foggy morning of January 20, 1909, about a mile and a half off the Lake Michigan shoreline at Seventy-First Street.

The city had built a temporary wooden crib to serve as a workstation and living quarters for 106 men who were digging and blasting a new tunnel under the lake. Farther out, at Sixty-Eighth Street, the tunnel would connect distribution facilities onshore with a new, permanent crib that had been finished the year before. As with the original crib built in the 1860s, the work was proceeding in both directions.

The men, mostly itinerant laborers, worked around the clock in twelve-hour shifts and lived in spartan quarters. About eight o'clock on this particular morning, some were sleeping in the second-floor bunk room, which was located directly above the dynamite storage area. Others were coming off duty, washing up and eating breakfast, while more were riding the bucket elevator down to the tunnel to begin another day.

Routine turned to emergency when black smoke began pouring from the powder storage room where at least two hundred pounds of dynamite were kept. (Some estimates placed the total much higher. An investigation failed to determine the exact amount.) Flames followed quickly, and within minutes, the wooden crib became a furnace. A desperate call to the contractor's field office was cut off mid-sentence when fire severed the phone line. Crew members of the contractor's service tug, tied up at the permanent crib half a mile away, spotted the smoke and steamed full speed ahead. The next sight that caught the sailors' eyes was that of men jumping off the crib—some with their clothes on fire—into the frigid water. Others were clinging to blocks of ice. Crewmen tossed life preservers, ropes, boxes—anything that would float—and pulled a number of victims into the tug. Then the boat maneuvered alongside the crib and more jumped in. The crew managed to save forty-eight workers, all suffering from burns, hypothermia or both. Nineteen crew members with less-severe injuries were dropped off at the Sixty-Eighth Street Crib to be picked up later. The tug took the remaining twenty-nine downtown to be transported to hospitals.

Two sets of victims never stood a chance: the fourteen who were sleeping in the bunk room above the dynamite storage area and were cremated and the twenty at the bottom of the shaft who either had been working there or descended in the futile hope of riding out the fire. All of those below died of smoke or gas inhalation. Some bodies were so badly burned that authorities couldn't determine the precise death toll. Seventy was the initial estimate, but it was later revised to sixty. Forty-six of the unidentified victims lie buried in a common grave at Mount Greenwood Cemetery on the Southwest Side. (The marker mistakenly lists forty-five men.)

A coroner's jury failed to determine the cause of the tragedy and discounted an early theory that it was caused by a worker who carried a lighted torch into the dynamite room. The jury further absolved the contractor and the city of any blame.

The temporary crib got rebuilt with wood and functioned for almost another three years, but it no longer housed any workers. The tunnel was completed in late 1911, ending the crib's usefulness. On November 6 of that year, the structure went up in flames again. This time, the fire was intentionally set as the most practical means of demolition. Some concerned,

Most victims of the 1909 temporary crib fire are buried in a common grave at Chicago's Mount Greenwood Cemetery. The fire was the worst disaster in the history of the city's water system. *From Mount Greenwood Cemetery Collection.*

but uninformed, nearby residents called the fire department to report that history was repeating itself.

> **1913** Damage to the Water Tower caused by the Great Fire of 1871 is finally repaired by using a combination of old and new limestone.
>
> **1914** W. Edbert Shoemaker and Florence Ann Brown are married on Sixty-Eighth Street Crib.

Shoemaker and Brown were swimming enthusiasts who met at the private Manhattan Beach (later the public Rainbow Beach) on the South Side, described by *Encyclopedia Chicago* as a popular gathering spot for middle-class boys and girls in the early twentieth century. The two began dating, then got engaged. They thought it logical, not to mention romantic, that the wedding venue have an aquatic theme. Some couples might have chosen a shipboard setting, a yacht club, something overlooking the lake or maybe a large tent on the beach. No, Ed and Florence wanted something more original. They opted for the Sixty-Eighth Street Crib. So, on an uncooperative August day with rolling waves and heavy mist, they set out from Manhattan Beach in a dory built by the groom. He manned the helm; she pulled a coat over her wedding dress to protect it from the spray. They were followed by a launch carrying parents and friends.

At the crib, a tender lowered a rope ladder, and everyone climbed up. The bride and groom took their positions on the footbridge above the water. The minister stepped forward, and the couple exchanged vows under leaden skies with screeching gulls circling overhead. Afterward, the wedding guests headed back to shore while the newlyweds set off on their honeymoon—a trip to St. Louis in their dory, which they had loaded with camping gear for overnight stays at riverside stopping points.

> **1918** Construction begins on the widening of Pine Street from the river to Oak Street to create the world-renowned boulevard North Michigan Avenue. A public outcry prevents city engineers from moving or razing the Water Tower to accommodate the merger of North Michigan with Lake Shore Drive.
>
> **1920** The ambitious widening project is completed with the opening of the Michigan Avenue Bridge, linking the

THE CHICAGO WATER TOWER

The Water Tower shows off a fresh look after a federally funded tuckpointing job in the mid-1930s. The Palmolive Building is to the left. *From Frank McMenamin collection.*

North and South Sides while creating the gateway to the Magnificent Mile.

1928 Floodlights illuminate the Water Tower at night.

1933 A civic ceremony at the tower posthumously honors waterworks pioneers Ellis Chesbrough, DeWitt Cregier and John Ernst Ericson with commemorative plaques.

1934 A force of more than seventy workers descends on the tower and Pumping Station to begin a two-month tuckpointing job. The cash-strapped city gets $35,000 from the Depression-era federal Works Progress Administration to complete the task.

1948 Once again, the Water Tower is a target for demolition. A group of art lovers petitions the city to tear down the tower and replace it with an art community center funded by $5 million in tax money. "Where would the city get $5 million?" aldermen want to know as Mayor Martin Kennelly vetoes the idea.

1963 The tower gets another facelift, this one costing $100,000. The refurbishing uses the last remaining yellow limestone from Joliet-area quarries—520 pieces. More than 700 old stones are reset, while future replacements are salvaged from a demolished church.

1967 Mayor Richard J. Daley and more than four hundred business and civic leaders and their guests gather at a North Michigan Avenue Hotel to kick off Water Tower Week in Chicago. The event marks the 100th anniversary of the tower's dedication.

1969 In honor of the 100th anniversary of its completion, the tower is named the first American water landmark by the American Water Works Association.

1971 Commemorating yet another centennial, that of the Great Fire, the city council designates the Water Tower as an official Chicago landmark. At long last, the preservation of the tower is assured.

1978 The 138-foot standpipe, whose concealment motivated construction of the tower, is removed, more than seventy years after being declared obsolete. A tourism center is added to the Chicago Avenue Pumping Station.

1981 Engine Company 98's firehouse becomes an official Chicago landmark.

1997 A tunnel leading from the Carter Harrison Crib collapses during an upgrade, snarling traffic on Lake Shore Drive and forcing the city to spend $5.3 million in repairs.

2001 The Pumping Station gets a new roof, new exterior lights and new landscaping, as well as other improvements.

2003 The Lookingglass Theatre opens in the Pumping Station's former boiler room.

2004 Engine Company 98's firehouse turns one hundred years old.

2014 Water Tower Park is renamed in honor of former Mayor Jane Byrne.

2017 The Harrison and Dever Cribs get a $5 million overhaul.

2019 Fifteen cultural organizations located near the Water Tower create a Water Tower Arts District to highlight their institutions and remind the public that there is more to the neighborhood than shopping and dining. And the Water Tower and Pumping Station celebrate their sesquicentennials.

Epilogue
AN ENDNOTE

Chicago historian Frank McMenamin combed his collection of postcards and shared ten images of the Water Tower for potential inclusion in this book. Four were written on and mailed, while the others remain blank. Postmarks range from 1906 to 1949—the cost of a stamp holding at one cent throughout. The unmarked cards were most likely popular within the latter part of this time span.

The common link among the images, beyond the obvious, is that each depiction, each treatment differs from its companions in background, perspective and, most notably, color (white, gray, beige, rust, brown). In two, the tower is barely visible within a skyline panorama. One is a night scene. Looking at the subtle differences, it's not difficult to understand how someone could sense that the tower itself had changed over the years.

Radio humorist Garrison Keillor used to amuse listeners by relaying mundane and frequently entertaining one-liners submitted by members of the live audience for acquaintances out there in Radioland. For example, "To Chris and Barb in Albuquerque—Powdermilk biscuits *are* mighty expeditious, and we're shipping some to you right after this broadcast. Love, Mom and Dad." Reading a stranger's postcard of long ago can generate that same feeling of eavesdropping on the ordinary. In January 1948, "Daddy" wrote to Charlotte in Alliance, Ohio, that it had been "a rather uneventful week…Getting colder tonight. Am tired and can't think of anything to say except I miss you—all of you. Write soon."

Epilogue

The Water Tower has been a ubiquitous image on Chicago postcards for generations. *From Frank McMenamin collection.*

Epilogue

OBSERVATORY TOWER AND WATER WORKS, CHICAGO. 97.

Epilogue

In June 1906, another Charlotte wrote, in German, to Wilhelmine in Milwaukee: "Greetings from Charlotte. We are arriving Sunday evening at eight or 10:30."

One of the cards segued from the pedestrian to the somber. Someone who might have signed "Uncle James"—the name is smudged—wrote in 1942 to Sarah in Columbus, Ohio. He mentioned that he had arrived in a snowstorm and was waiting for a bus to Freeport (Illinois, presumably). Then he added, "Have read your letter over and over again. How profoundly you have expressed that the hand of God is leading you in this time of crisis. He hast said, 'I will never leave or forsake thee.' With much love,…"

The reader can only wonder about Sarah's crisis and hope that she survived it. Or that Dad made it back home safely to Alliance, that Charlotte and Wilhelmine got together in Milwaukee, Nate found Ben in Boston after the former's trip to LA. (Nate liked the weather in Chicago on March 8, 1949.) These simple messages stimulate our curiosity about how things turned out. It's like reading a blurb without being able to get into the book. Who were these people, and how did their postcards end up in various flea markets decades later? Were their lives happy or sad or, more likely, a combination of both? All we know is that for one moment they reached out to someone who meant something to them on the back of a Chicago Water Tower penny postcard.

BIBLIOGRAPHY

Andreas, A.T. *History of Chicago.* Chicago: self-published, 1884.
Beatty, William K. "When Cholera Scourged Chicago." *Chicago History* 11, no. 1 (Spring 1982).
Cain, Louis P. "Raising and Watering a City: Ellis Sylvester Chesbrough and Chicago's First Sanitation System." *Technology and Culture* 13, no. 3 (1972).
Chesbrough, Ellis S. *Chicago Sewerage Report.* Chicago: Board of Sewerage Commissioners, 1858.
Chicago Historical Society. "The Masons and the Water Works." *Chicago History* (Fall 1954–Summer 1957).
City of Chicago. *Old Chicago Water Tower District.* Chicago Municipal Reference Library, 1984.
———. "Pictorial Story of Chicago Public Works." *Civil Service News*, 1937.
Currey, J. Seymour. *Chicago: Its History and Its Builders.* Chicago: S.J. Clarke, 1912.
Drury, John. *Old Chicago Houses.* Chicago: University of Chicago Press, 1975.
Gale, Edwin O. *Reminiscences of Early Chicago and Vicinity.* Chicago: Fleming H. Revell Company, 1902.
Hall, Jennie. *The Story of Chicago.* Chicago: Rand McNally, 1911.
Isaacson, Walter. *Leonardo Da Vinci.* New York: Simon and Schuster, 2017.
Kirkland, Joseph. *History of Chicago Illinois, Volume II.* Chicago: Munsell, 1895.
Lanyon, Richard. *Draining Chicago: The Early City and the North Area.* Chicago: Lake Claremont Press, 2016.

Bibliography

Little, Ken, and John McNalis. *History of Chicago Firehouses, Volume II*. Chicago: self-published, 2000.

Mayer, Harold M., and Richard C. Wade. *Chicago: Growth of a Metropolis*. Chicago: University of Chicago Press, 1969.

McMenamin, Frank J. *The Fight for Chicago: The Great Chicago Fire October 7 and 8, 1871*. Chicago: self-published, n.d.

Miller, Donald C. *City of the Century: The Epic of Chicago and the Making of America*. New York: Simon and Schuster, 1996.

Nelson, Evelyn S. "History of the Water Tower: Chicago's Famous Landmark." Unpublished manuscript, 1961.

"Palmolive Building Beacon." *Modern Mechanics and Inventions*, 1930.

Pierce, Bessie Louise. *A History of Chicago, Volume II, from Town to City 1848–1871*. New York: Alfred A. Knopf, 1940.

Stamper, John W. *Chicago's North Michigan Avenue: Planning and Development, 1900–1930*. Chicago: University of Chicago Press, 2005.

Tallmadge, Thomas E. *Architecture in Old Chicago*. Chicago: University of Chicago Press, 1941.

Wing, Jack. *The Great Chicago Lake Tunnel*. Chicago: Western News Company, 1867.

INDEX

A

Allerton Hotel 83
American Water Works Association 17

B

Berg, Anton 31
Board of Public Works 43, 53, 69, 108
Board of Sewerage Commissioners 37, 107
Boyington, William Warren 52
Bridgeport 40, 61
Bross, William 31, 32
Burnham, Daniel 61
Byrne, Jane 15, 16, 97, 115
Byrne, Kathy 15, 16

C

Carter H. Harrison Crib/William E. Dever Crib 103
Chesbrough, Ellis Sylvester 37, 38, 39, 40, 43, 44, 50, 52, 53, 55, 56, 57, 58, 61, 62, 73, 99, 102, 103, 107, 108, 113
Chicago Avenue Pumping Station 23, 90, 110, 114
Chicago Board of Trade Building 53
Chicago Fire Department's Engine Company 98 and Ambulance 11 29
Chicago Fire Department's Training Academy 63
Chicago Firehouse Restaurant 98
Chicago Historical Society 82
Chicago Hydraulic Company 32, 106
Chicago Lake Tunnel 45
Chicago River 31, 32, 35, 40, 61, 64, 109

Index

Chicago's Rosehill Cemetery 52
Chicago Water Works 21
cholera 35, 36
city council 15
City Gallery 19
Cobb, Henry Ives 78
Coglianese, Lieutenant Edmond 95, 97
courthouse 48, 50, 56, 65
Cregier, DeWitt Clinton 17, 53, 56, 57, 60, 69, 73, 113
Cyrus McCormick residence 76

D

Daley, Mayor Richard J. 16, 85, 98
Department of Water Management 102
Diversey and Lill Brewery 67
double house 76
Drake Hotel 76
Dull and Gowan 44
Dunne, Edward F. 99

E

Engine Company 98 94
Eugene Sawyer Plant 100

F

Farwell brothers, John V. and Charles B. 76
Florence Cathedral 52
Foster, Dr. John H. 71
Four-Mile Crib 102

Fourth Presbyterian Church 82
Frost, Charles Sumner 78
Fuller, D.W. 67
Fuller, S.W. 67

G

Gold Coast 79, 93
Gorin, Jerome R. 57
Great Chicago Fire of 1871 23, 59, 63
Gurnee, Mayor Walter 33

H

Harrison II, Mayor Carter 72
Hegeler-Carus Mansion 53
Hermann, E.F. 94
hotel 42, 49, 78, 83, 92, 93, 114

I

icebreaker *James J. Versluis* 100
Illinois and Michigan Canal 61
Illinois Women's Athletic Club 87

J

James W. Jardine Water Purification Plant 100
Jane Byrne Park 16
John Hancock Observatory 21
Joseph T. Ryerson family residence 75

Index

K

Keillor, Garrison 117
Kirkland, Joseph 35

L

Lake Michigan 15, 31, 35, 45, 74, 90, 103, 110
Lake Shore Drive 60, 82, 93, 104, 112, 114
Lake Shore Park 60
Lake View Crib 104
LaSalle Street financial district 65
La Salle Street Station 53
Lawrence Avenue Crib 104
Lewis Tower 88
Lindbergh Beacon 89
Little, Ken 93
Lookingglass Theatre 24

M

Magnificent Mile, the 82
Mahlon Ogden residence 70
Mark Twain Hotel 95
Masonic Fraternity 19
McAlpine, William J. 33
McCormick, Edith Rockefeller 76
McCormick, Harold 76
McMenamin, Frank 117
McVickers Theater 65
Morgan Iron Works 53
Museum of Contemporary Art 60

N

Navy Pier 100
Newberry Library 70
900 North Michigan Building 76
North Pier 47
North Western Railroad Depot 67

O

Ogden, William Butler 37
Oglesby, Illinois governor Richard 47
Old Sallis 24
Old Sally 33, 53, 60, 70, 73, 107
old state prison at Joliet 52
O'Leary, Catherine 63
O'Leary, Patrick 63

P

Palmer House 65, 78
Palmer, Potter and Bertha 78
Palmolive Building 89
Perry Smith House 78
Pine Street 59, 61, 67, 74, 75, 79, 80, 81, 82, 93, 112
Plan of Chicago 80
Playboy Enterprises 89
Preservation Chicago 102
Pullman, George Mortimer 42

INDEX

R

Rainbow (Seventy-Ninth Street) Beach 100
Ritz-Carlton Hotel 92
Rockefeller, John D. 76

S

Saint James Episcopal Church 65
Saint Paul's Catholic Church 65
Second Presbyterian Church 53
Sherman House Hotel 53
Sherman, Mayor Francis 45
Sixty-Eighth Street Crib/Edward F. Dunne Crib 103

T

Thompson, Mayor William H. 81
Tip-Top Tap 85
Trautman, Frank 67, 68, 69
Tree, Judge Lambert 65
Tremont House 42
Two-Mile Crib 99

U

Union Pacific and Central Pacific 50

W

Washington Square Park 71
Water Tower Place 17, 74, 92
Wilde, Oscar 15
William E. Dever Crib 99
Williams, Robert 64
Willis (Sears) Tower 90
Woman's Chicago Athletic Club 88
Wright's Livery Stable 67

ABOUT THE AUTHOR

Chicago native John F. Hogan is a published historian and former broadcast journalist and on-air reporter (WGN-TV/Radio) who has written and produced newscasts and documentaries specializing in politics, government, the courts and the environment. As WGN-TV's environmental editor, he became the first recipient of the United States Environmental Protection Agency's Environmental Quality Award. His work also has been honored by the Associated Press. Hogan left broadcasting to become director of media relations and employee communications for Commonwealth Edison Company, one of the nation's largest electric utilities. Hogan is the author of Edison's one-hundred-year history, *A Spirit Capable*, as well as five other Chicago books with The History Press: *Chicago Shakedown*, *Fire Strikes the Chicago Stock Yards*, *Forgotten Fires of Chicago*, *The 1937 Chicago Steel Strike* and *The Great Chicago Beer Riot*. He holds a Bachelor of Science in journalism/communications from the University of Illinois at Urbana–Champaign and presently works as a freelance writer and public relations consultant.

Visit us at
www.historypress.com